Rock Mechanics Through Project Based Learning

Rock Mechanics Through Project-Based Learning

Ivan Gratchev

Griffith School of Engineering, Gold Coast, Southport, QLD, Australia

CRC Press is an imprint of the
Taylor & Francis Group, an **informa** business
A BALKEMA BOOK

CRC Press/Balkema is an imprint of the Taylor & Francis Group, an informa business

Typeset by Apex CoVantage, LLC

Library of Congress Cataloging-in-Publication Data

Applied for

Published by: CRC Press/Balkema
Schipholweg 107C, 2316 XC Leiden, The Netherlands
e-mail: Pub.NL@taylorandfrancis.com
www.crcpress.com – www.taylorandfrancis.com

ISBN: 978-0-367-23219-1 (Pbk)
ISBN: 978-0-367-23203-0 (Hbk)
ISBN: 978-0-429-27883-9 (eBook)

DOI: https://doi.org/10.1201/9780429278839

Contents

Preface

Our experience as teachers indicates that a traditional format where the instructor provides students with theoretical knowledge through a series of lectures and abstract textbook problems is not sufficient to prepare students to tackle real geotechnical challenges. There is a disjunct between what students are taught in universities and what they are expected to do as engineers. Although students may easily derive complex equations in class, they often struggle when faced with practical engineering problems in the workplace. It is thus not surprising that a project-based approach to learning, which partially simulates what engineers do in real life, has proved to be a valid alternative to the traditional method (Gratchev and Jeng, 2018; Gratchev et al., 2018). Project-based learning not only provides students with opportunities to better understand the fundamentals of geotechnical engineering, but it also allows them to gain more practical experience and learn how to apply theory to practice. In addition, working on a practical project makes the learning process more relevant and engaging.

To our knowledge, there is currently no textbook that utilizes a project-based approach to introduce theoretical aspects of rock mechanics to students. This book will appeal to new generations of students who would like to have a better idea of what to expect in their employment future. In this book, readers are presented with a real-world challenge (in the form of a project-based assignment) similar to that which they would encounter in engineering practice and they need to work out solutions using the relative theoretical concepts which are briefly summarized in the book chapters. To complete this project-based assignment, readers are required to undertake a sequence of major tasks including (a) interpretation of field and laboratory data; (b) analysis of rock properties and rock mass conditions, (c) identification of potential rock-related problems at a construction site and (d) assessment of their effect on slope stability and tunnel construction.

This book covers all significant topics in rock mechanics and discusses practical aspects related to rock falls, rock slope stability and tunnels. Each section is followed by several review questions that will reinforce the reader's knowledge and make the learning process more engaging. A few typical problems are discussed at the ends of chapters to help the reader develop problem-solving skills. Once the reader has sufficient knowledge of rock mechanics, they will be able to undertake a project-based assignment to scaffold their learning. The assignment is based on real field and laboratory data including boreholes and test results so that the reader can experience what engineering practice is like, identify with it personally and integrate it into their own knowledge base. In addition, some problems will include open-ended questions, which will encourage the reader to exercise their judgement and develop practical skills. To foster the learning process, solutions to all questions will be provided and discussed.

The author is grateful to all students of the Rock Mechanics course at Griffith University for their constructive feedback in the past several years. This book would not have been the same without their enthusiasm and interest in geotechnical engineering. The support from Ekaterina Gratchev is gratefully acknowledged as well.

Conversion factors

Length	*Mass and weight*	*Area*	*Volume*	*Unit weight*	*Stress*
1 in = 2.54 cm 1 ft = 30.5 cm	1 lb = 454 g 1 lb = 4.46 N 1 lb = 0.4536 kgf	1 in^2 = 6.45 cm^2 1 ft^2 = 0.0929 m^2	1 ml = 1 cm^3 1 l = 1000 cm^3 1 ft^3 = 0.0283 m^3 1 in^3 = 16.4 cm^3	1 lb/ft^3 = 0.157 kN/m^3	1 lb/in^2 = 6.895 kN/m^2 1 lb/ft^2 = 47.88 N/m^2
1 m = 39.37 in 1 m = 3.281 ft	1 N = 0.2248 lb 1 metric ton = 2204.6 lb 1 kgf = 2.2046 lb	1 m^2 = 10.764 ft^2 1 cm^2 = 0.155 in^2	1 m^3 = 35.32 ft^3 1 cm^3 = 0.061023 in^3	1 kN/m^3 = 6.361 lb/ft^3	1 kN/m^2 = 20.885 lb/ft^2 1 kN/m^2 = 0.145 lb/in^2

About the author

Ivan Gratchev is a senior lecturer at the Griffith School of Engineering, Griffith University, Australia. He graduated from Moscow State University (Russia) and received his PhD from Kyoto University (Japan). He worked as a research fellow in the geotechnical laboratory of the University of Tokyo (Japan) before joining Griffith University (Australia) as a lecturer. His research interests are in geotechnical aspects of landslides, soil liquefaction and rock mechanics. He has published numerous research articles in leading international journals and international conferences.

Since joining Griffith University in 2010, he has taught several geotechnical courses (including soil mechanics, rock mechanics and geotechnical engineering practice) using a project-based approach. His teaching achievements were recognized by his peers and students through a number of learning and teaching citations and awards.

Chapter 1

Introduction and book organization

1.1 Rocks and rock mechanics

This book is about rocks and their properties. We often use the word 'rock' in our daily life, but does it have the same meaning in rock mechanics? Commonly used definitions state that the term 'rock' refers to a hard substance made of minerals, which requires drilling, blasting, wedging, or other brute force to excavate. Many engineering structures are built on rocks, while rocks are commonly used as construction material for several engineering applications. For this reason, engineers and researchers are required to have a sound knowledge of rock properties and understand the behavior of rock mass under stresses.

Question: *Concrete and dry clay are hard as well; are they also rock?*
Answer: No, dry clay and concrete are not rock. Although concrete is as hard as rock, it is artificial material, while clay becomes very soft when saturated.

Rock mechanics is the subject that deals with rock response to applied disturbance caused by natural and engineering processes. Through this book, the reader will not only learn the fundamentals of rock mechanics, but they will also see practical applications of the theoretical knowledge.

Question: *Is rock mechanics the same as engineering geology?*
Answer: No, they are different. Engineering geology mostly deals with the application of geological fundamentals to engineering practice, while rock mechanics covers the engineering properties of rock.

1.2 Book organization

The book is organized in such a way that each chapter first explains its relevance to the project and then it briefly provides key theoretical concepts necessary to complete a certain part of the project. Chapter 2 provides the project description and data from field investigation and laboratory testing. Chapters 3 and 4 are related to basic geology as they deal with the effects of geology on rock formation (Chapter 3) and common types of rock (Chapter 4). Chapter 5 discusses common rock exploration methods and techniques, while Chapter 6 is dedicated to discontinuities in rock mass. Rock properties and rock testing methods are described in Chapters 7 and 8, respectively. Chapters 9–12 show practical applications of rock mechanics, including the assessment of rock mass properties (Chapter 9), rock fall (Chapter 10) and landslide (Chapter 11) disaster, and common issues with rock mass during tunnel construction (Chapter 12).

Each chapter also provides a few practical problems that the reader can use for more practice. It is suggested to solve each problem first before referring to the step-by-step solution provided afterwards. Even though it may be difficult to work out the final answer, spending time on each problem will improve the reader's understanding of the relevant material and help to develop problem-solving skills. To reinforce the knowledge of rock behavior and review the key concepts, the reader can also take a review quiz at the end of each chapter.

Chapter 2

Project description

This project deals with a dam built in Southeast Asia in 1970s. After its completion, there was a concern that the neighboring slopes might experience stability issues during rainy seasons. The risk analysis indicated that if the slope failed, it would affect the major transportation routes around the dam. To collect more information about the slope conditions, field investigations including geological mapping, boreholes and joint surveys were performed. A series of laboratory tests were conducted on core specimens to obtain the engineering properties of rock. The field and laboratory data is given in the following sections.

2.1 Data from site investigation

Figure 2.1 presents a map of the studied area. Four boreholes were drilled along the A-A′ line, and the borehole logs are given in Figures 2.2–2.5.

2.2 Data from laboratory testing

Initial examination of core size and its mass was performed to determine rock density. The obtained data for three rock types – sandstone, mudstone and andesite – is given in Table 2.1.

For each type of rock, laboratory testing was conducted to determine its engineering properties. Intact specimens of sandstone collected at a depth of 2 m were used for a series of triaxial tests. The lab data from these tests is summarized in Table 2.2 in terms of the principal stresses (σ_{1f}, σ_{3f}) at failure.

The samples of mudstone were collected at a depth of 7 m for a series of point load tests with an axial direction of load. The sample size and the force recorded at failure are given in Table 2.3.

A cylindrical specimen of andesite (diameter of 50 mm and height of 100 mm) collected at a depth of 18 m was tested under unconfined compression and the obtained results are presented in Table 2.4.

A few other tests performed on rock specimens will be discussed later in this book, including slake durability and Schmidt hammer tests (Chapter 7).

2.3 Project tasks

The objectives of this project are to establish the geological setup of the site, determine the rock mass properties, estimate slope stability along the A-A′ line (Figure 2.1), perform rock fall hazard assessment (at Point R in Figure 2.1) and discuss geotechnical issues that may

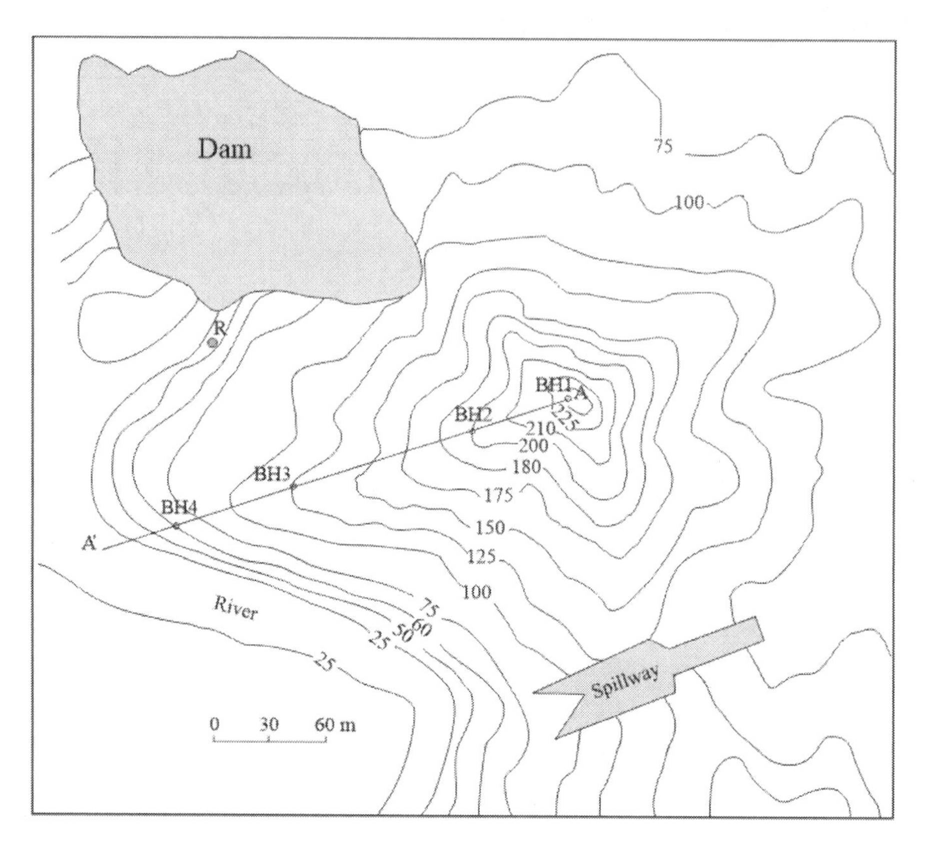

Figure 2.1 Map of the studied area

occur during tunnel construction. The project includes several tasks which are connected as follows:

1. Analysis of geological conditions at the project site. This involves determining the major geological units, identifying geological structures and discussing their effect on rock properties (Chapters 3 and 4).
2. Analysis of the borehole logs and drawing a cross-section along the A-A′ line. This involves describing the properties of each geological unit and identifying the rock layer(s) that may cause geotechnical issues. This part will be explained in Chapter 5.
3. Analysis of joint characteristics of the rock mass and their effect of rock strength will be discussed in Chapter 6.
4. Analysis of the lab data and determination of the engineering properties of intact rock specimens. This involves estimating the rock properties such as porosity, compressive and tensile strength, cohesion and friction angle. The laboratory procedures and data interpretation techniques will be discussed in Chapters 7 and 8.

Drilling Method	Water	RQD	Depth (m)	Graphic Log	Material Description	Weathering	Strength	Defect description	Elevation (m)
NMLC		0 19 33	0–3		SANDSTONE Heavily weathered, moist and weak sandstone, grey, with some fine bands of dark grey shale to 20 mm.	EW DW	VL L	Joint are slickensided, undulating, and highly weathered; joints are separated by about 3-5 mm and filled with clay.	230 229 228
		71 84	3–10		MUDSTONE Slightly weathered, dark, massive mudstone	SW F	M H	Joint surfaces are slightly rough, slightly weathered with stains; no clay found on surface, apertures generally less than 1 mm.	227 226 225 224 223 222 221
		89 92 94	10–20		ANDESITE Very hard, dark bedrock	SW-F F	VH	Joint surfaces are generally stepped and rough, tightly closed and unweathered with occasional stains.	220 219 218 217 216 215 214 213 212 211
			20		End of borehole				

Figure 2.2 Borehole BH1, Elevation – 231 m. Borehole log legend: Drilling method: NMLC – diamond core 52 mm. Weathering: EW – extremely weathering, DW – distinctly weathering, SW – slightly weathering, F – fresh. Strength: VL – very low, L – low, M – medium, H – high, VH – very high.

Drilling Method	Water	RQD	Depth (m)	Graphic Log	Material Description	Weathering	Strength	Defect description	Elevation (m)
NMLC		0 23 36	1 2 3 4		SANDSTONE Heavily weathered, moist and weak sandstone, grey, with some fine bands of dark grey shale to 20 mm.	EW DW	VL L M	Joint are slickensided, undulating, and highly weathered; joints are separated by about 3-5 mm and filled with clay.	199 198 197 196
		75 84	5 6 7 8 9 10 11 12		MUDSTONE Slightly weathered, dark, massive mudstone	SW F	H	Joint surfaces are slightly rough, slightly weathered with stains; no clay found on surface, apertures generally less than 1 mm.	195 194 193 192 191 190 189 188
		91 92 94	13 14 15 16 17 18 19 20		ANDESITE Very hard, dark bedrock	SW-F F	VH	Joint surfaces are generally stepped and rough, tightly closed and unweathered with occasional stains.	187 186 185 184 183 182 181 180
					End of borehole				

Figure 2.3 Borehole BH2, Elevation – 200 m. Borehole log legend: Drilling method: NMLC – diamond core 52 mm. Weathering: EW – extremely weathering, DW – distinctly weathering, SW – slightly weathering, F – fresh. Strength: VL – very low, L – low, M – medium, H – high, VH – very high.

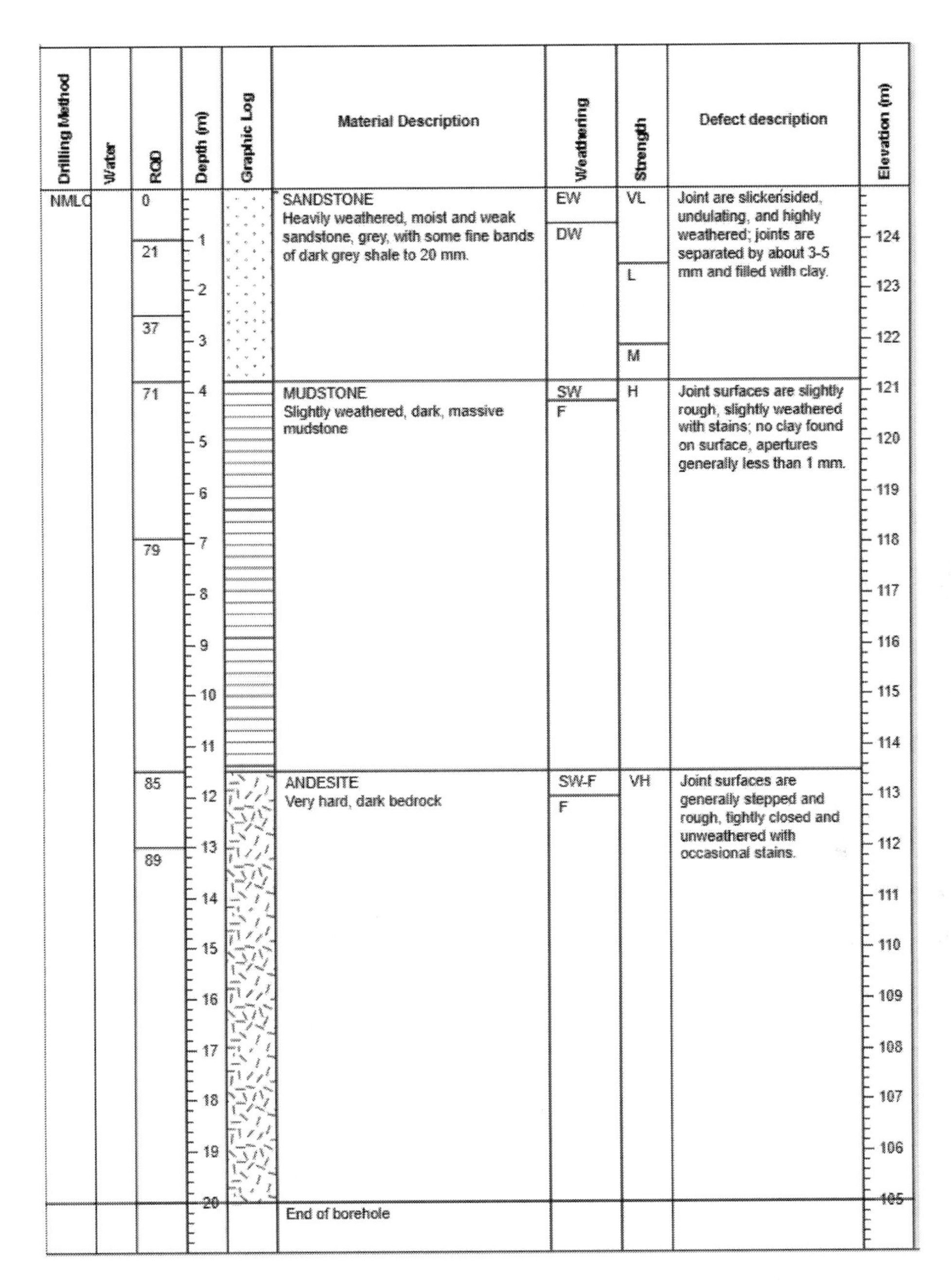

Figure 2.4 Borehole BH3, Elevation – 125 m. Borehole log legend: Drilling method: NMLC – diamond core 52 mm. Weathering: EW – extremely weathering, DW – distinctly weathering, SW – slightly weathering, F – fresh. Strength: VL – very low, L – low, M – medium, H – high, VH – very high.

Drilling Method	Water	RQD	Depth (m)	Graphic Log	Material Description	Weathering	Strength	Defect description	Elevation (m)
NMLC		0 33	1 2		SANDSTONE Heavily weathered, moist and weak sandstone, grey, with some fine bands of dark grey shale to 20 mm.	EW DW	VL L	Joint are slickensided, undulating, and highly weathered; joints are separated by about 3-5 mm and filled with clay.	59 58
		71 84	3 4 5 6 7		MUDSTONE Slightly weathered, dark, massive mudstone	SW F	M H	Joint surfaces are slightly rough, slightly weathered with stains; no clay found on surface, apertures generally less than 1 mm.	57 56 55 54 53
		89 94	8 9 10 11 12 13 14 15 16 17 18 19		ANDESITE Very hard, dark bedrock	SW F	VH	Joint surfaces are generally stepped and rough, tightly closed and unweathered with occasional stains.	52 51 50 49 48 47 46 45 44 43 42 41
			20		End of borehole				40

Figure 2.5 Borehole BH4, Elevation – 60 m. Borehole log legend: Drilling method: NMLC – diamond core 52 mm. Weathering: EW – extremely weathering, DW – distinctly weathering, SW – slightly weathering, F – fresh. Strength: VL – very low, L – low, M – medium, H – high, VH – very high.

Table 2.1 Initial examination of core samples

Mass (g)	*Diameter (mm)*	*Height (mm)*
	Sandstone	
470	51.3	101.2
456	51.9	101.6
469	51.2	101.1
	Mudstone	
440	50.9	100.9
426	51.1	101.1
454	51.2	101.3
	Andesite	
540	50.2	101.8
522	50.5	101.6
551	50.6	101.8

Table 2.2 Results of triaxial tests on samples of sandstone

σ_{3f} (MPa)	0	5	10	15	20	25
σ_{1f} (MPa)	25	50	72	90	105	120

Table 2.3 Results from a series of point load tests on mudstone

No.	*Rock size*		*Load at failure*
	W	*D*	*P*
	mm	*mm*	*kN*
1	50.9	37.2	5.0
2	51.2	35.3	4.1
3	50.6	33.6	3.9

Table 2.4 Results from an unconfined compression test on andesite

Load (kN)	*Strain (%)*
0	0.000
26	0.013
49	0.025
68	0.038
85	0.050
100	0.063
111	0.075
116	0.088
118	0.100
110	0.105

5. Determination of rock mass ratings. Using the data from tasks 3 and 4, it will be possible to determine the rock mass ratings such as RMR (Rock Mass Rating) and GSI (Geological Strength Index) (Chapter 9).
6. Analysis of the rock mass properties. The data from tasks 3 and 5 will be combined to determine the strength characteristics of the rock mass (Chapter 9).

The rock mass properties determined after the completion of the first six tasks will be used to solve practical problems such as slope stability (Chapter 11) and engineering issues during tunnel construction (Chapter 12). Chapter 10 will introduce and discuss the principles of rock fall hazard assessment.

Chapter 3

Rock mass formation

Project relevance: During geological time, rocks are affected by natural processes including tectonic movements, which result in the formation of various geological structures. It is important to become familiar with commonly occurring geological structures and have a good understanding of their effect on rock mass properties. This chapter will introduce basic geology and discuss the effect of rock structures in relation to rock properties.

3.1 The structure of the Earth and tectonic activities

We will start with basic geology and investigate the composition of the Earth.

Question: *Why do engineers need to study geology? How is it related to rock mechanics?*
Answer: It is important to know geology and the effect of geological processes on the formation of rock mass because this knowledge will help us better understand the response of rock to applied stresses during construction.

The Earth is composed of three components: crust, mantle and core (Figure 3.1). The crust, which has an average thickness of 30–35 km in continents and about 5 km in oceans, is of most interest to engineers as this is where all infrastructure (including deep mines) is built. The lithosphere, which has a thickness of about 100 km, consists of the crust and the outer (solid) part of the mantle (Figure 3.1). The asthenosphere is the upper mantle which is made of softer (melt) material, and it is located below the lithosphere.

Rock formation is a complex process that occurs in the lithosphere, and it is accompanied by tectonic activities such as plate movements. According to plate theory, the lithosphere is broken into plates, which slowly move above the asthenosphere (Figure 3.2). The collision (Figure 3.2a) or subduction (Figure 3.2b) of plates results in the formation of ridges or trenches, respectively, accompanied by volcanism and earthquakes. These tectonic movements can (a) rupture rock mass, forming faults and (b) deform rock mass by creating folds. In both cases, the engineering properties of rock mass will alter (that is, the rock mass will become jointed and more susceptible to weathering).

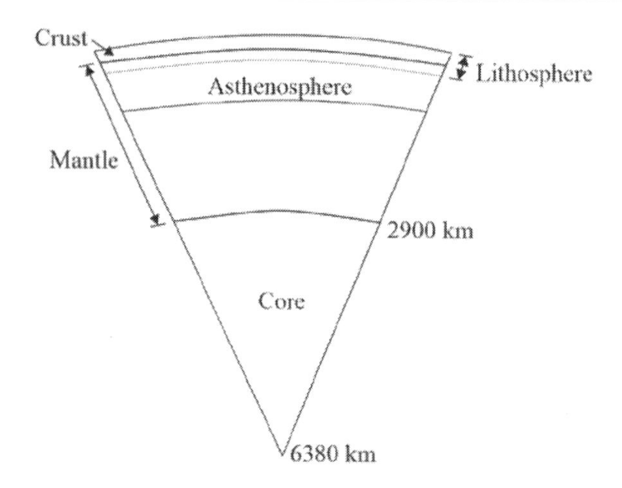

Figure 3.1 Structure of the earth (not to scale)

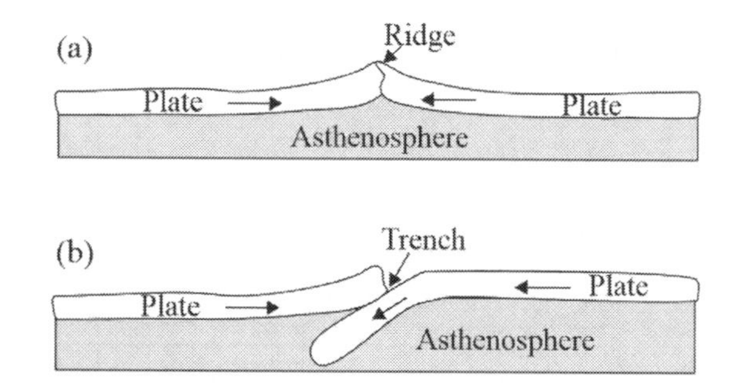

Figure 3.2 Tectonic plate movements resulting in the formation of ridges (a) and trenches (b)

Question: *What is the typical rate of plate movement?*
Answer: It depends on the plate and location. For example, the Indo-Australian Plate in the eastern part of Australia moves at the rate of about 5 cm per year, while the western part (India) moves at the rate of less than 4 cm per year due to the impediment of the Himalayas.

3.2 Geological structures

Tectonic movements lead to the formation of geologic structures such as faults, folds and shear. Such structures need to be identified during site investigations because they are associated with fractured rocks and potential engineering issues.

3.2.1 Faults

Geological faults are planar rock fractures that show evidence of relative movement (Figure 3.3). As faults typically do not consist of a single, clean fracture, the term *fault zone* is commonly used for the zone of complex deformation associated with the fault plane.

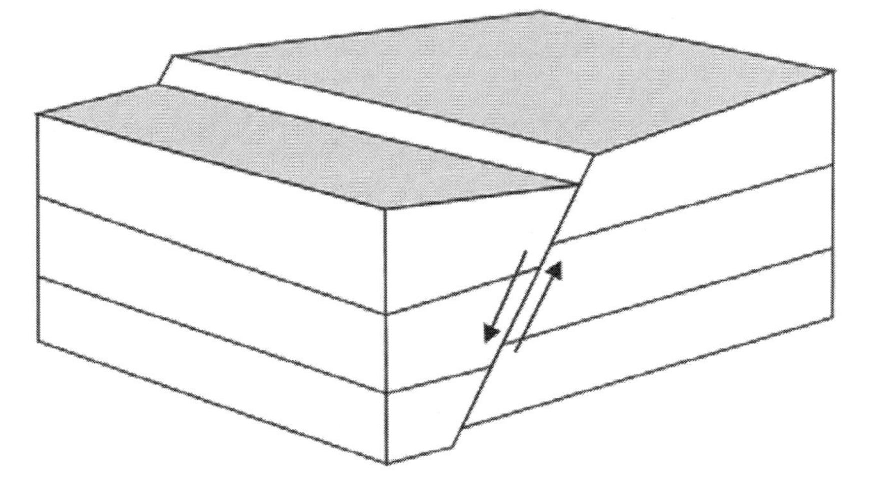

Figure 3.3 A schematic diagram of faults

Figure 3.4 Tectonic fault that goes underneath the bridge foundation. As a result, the bridge experienced significant damage during an earthquake.

Minor faults may have offsets measured in millimeters, while major faults can have shear displacements measured in kilometers.

Faults are associated with past or recent tectonic activities, and engineering structures built in the vicinity of faults can experience significant deformation (Gratchev et al., 2011) during earthquakes (Figure 3.4).

Faults can be found in non-seismically active areas as well. Such faults are usually not active but they can still cause some engineering issues. Rocks along non-active faults can still be heavily fractured, making it rather difficult to excavate due to slope stability concerns (Figure 3.5).

Question: *What is an active fault?*
Answer: An active fault is a fault that will move again. This means that there is a reasonable chance of slip occurring during the design life of an engineering structure, which can be about 100 years for a dam (Sherard et al., 1974). Faults that do not meet this criterion may be called dormant (non-active).

3.2.2 Folds

Folds (Figure 3.6) are produced when rock layers are deformed to give a waved surface, as schematically shown in Figure 3.7. They are typically caused by tectonic forces including the horizontal compression resulted from the gradual cooling of the Earth's crust. Folds can be of different scales, from meters to kilometers.

They are commonly found in the beds of sedimentary rocks and often associated with a higher degree of rock fracturing. Fracturing may be intensified along the loci of highest curvature, which may catalyze weathering in those zones.

Question: *From an engineering point of view, what is the effect of faulting and folding on the mechanical defects of rocks?*

Figure 3.5 The quarry wall consists of heavily fractured rock, which makes it rather difficult to excavate due to slope stability issues

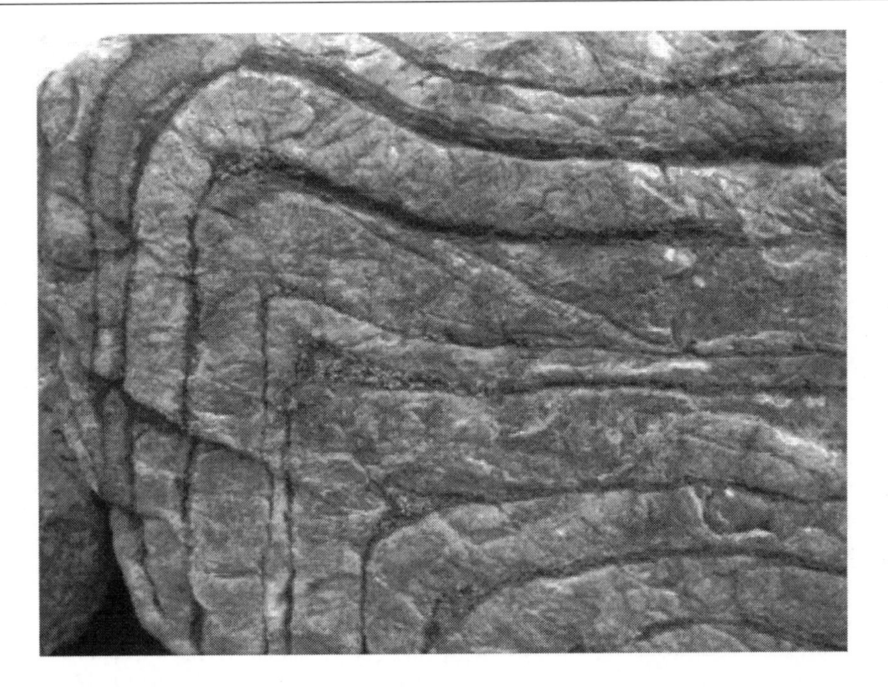

Figure 3.6 Folds in rock

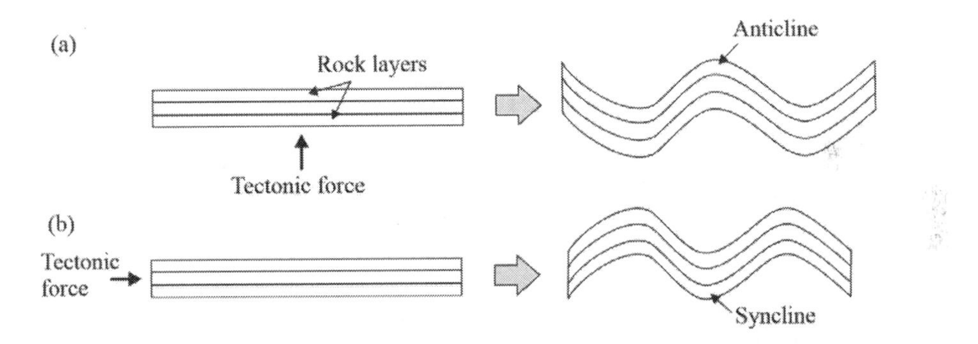

Figure 3.7 The role of tectonic forces in the formation of folds in rock mass. Rock layers undergo deformations under vertical (a) and horizontal (b) forces. Anticlines and synclines are the up and down folds that usually occur together and are caused by compressional stress.

Answer: Faults and folds are typically associated with fractured rocks. The mechanical rock defects depend on the stress-deformation characteristics of the rock. Rocks which are sufficiently strong to transmit compressive force under given conditions are considered to be competent under these conditions (e.g., igneous rocks and sandstone). Rocks such as shale and slate, which are sufficiently plastic to deform without fracturing, are incompetent. Thus competent strata in folded regions are likely to be intensely fractured while incompetent strata in the same region may be almost or entirely intact (Terzaghi, 1946). The degree of competence also determines the rock mass conditions in the vicinity of faults; that is, if the rock is incompetent, the fault may be barely visible, while in competent strata, the rock may be completely crushed.

3.2.3 Bedding planes

Bedding planes divide rocks into beds or strata; they represent interruption in the sedimentary process or repeated sedimentation cycles. Bedding planes can exist in one rock type (e.g., sandstone) or be an alternation of two or three types of rock (e.g., sandstone and argillite, Figure 3.8). When two rocks form bedding planes, such a geological structure is also referred to as *flysch*. It is noted that over time, bedding planes can become potential weathered zones and contain pockets of ground water.

Question: *Why are sedimentary rocks commonly stratified?*
Answer: Sedimentary rocks are formed from weathered rock material such as gravel, sand, silt and/or clay. This weathered material is transported to the sedimentation zone (oceans and seas) by brooks and rivers. The velocity of the current that transports sediments varies with the seasons. At low velocity the water can carry only very small particles, whereas at high velocities coarse particles can also be carried in suspension (Proctor and White, 1946). Therefore, layers of sandstone (which are made of coarse sediments) can be separated from each other by thin layers of shale (which are made of fine sediments), while beds of coarse-grained sandstone may alternate with beds of fine-grained sandstone.

Figure 3.8 Bedding planes exposed by a road cut

Question: *Is mélange the same thing as bedding?*
Answer: No, mélange is a large-scale breccia that lacks continuous bedding. It includes fragments of rock of all sizes, which are contained in a fine-grained, deformed matrix. Mélange is typically found in the area of plate subduction and it is formed by tectonic movements of the lithosphere.

3.2.4 Shear zones

A shear zone (Figure 3.9), which varies from a few centimeters to several meters, is commonly associated with structural instabilities such as *landslides*. The shear zone can be filled with fine material such as silt or clay, making it even weaker during rainfall as the infill becomes saturated.

3.3 Rock weathering

Over geologic time, rocks are exposed to weathering, a process that transforms hard rocks into soft soil. It is important for engineers to learn how rocks respond to surface weathering because most of civil work intersects rock at shallow depths where rocks have likely undergone significant changes.

Question: *When you mention 'geologic time', how is it different from 'normal time'?*
Answer: Geologic time counts millions of years and it is divided into time units called eras and subdivided further into periods and epochs (Table 3.1). We should not ignore the age of rocks because it gives important information about the rock properties. For example,

Figure 3.9 A view of the shear zone that crosses the tunnel's roof

Table 3.1 Geologic time and divisions (after Verhoogen et al., 1970)

Era	*Period*	*Epoch*	*Absolute age (millions of years)*
Cenozoic	Quaternary	Holocene	
		Pleistocene	3
	Tertiary	Pliocene	
		Miocene	70
		Oligocene	
		Eocene	
		Paleocene	
Mesozoic	Cretaceous		
	Jurassic		230
	Triassic		
Paleozoic	Permian		
	Carboniferous	Pennsylvanian	600
		Mississippian	
	Devonian		
	Silurian		
	Ordovician		
	Cambrian		
Precambrian			Earlier

the older Precambrian rocks (relatively old rocks) tend to be very hard, crystalline material with many fractures, while some Pliocene sandstones (relatively young rocks) can be as porous as soil.

Rocks become more porous, permeable and weak and eventually disintegrate into soil as a result of mechanical (also called 'physical') and chemical weathering. Mechanical weathering (e.g., heating and cooling of rocks) is the breakdown of rocks into particles without producing changes in the chemical composition of the rock (Gratchev et al., 2019). Chemical weathering is the breakdown of rock by chemical reactions, and water plays a crucial role in this process. Rainwater often initiates chemical weathering because of the chemical reactions that occur between atmospheric gases dissolved in water and rocks. Chemically weathered rocks are discolored and have stains on their surfaces. As this process continues, the rock minerals break into smaller particles, which can be easily carried away by water or wind.

Question: *Does the weathering rate depend on location?*
Answer: It depends more on climate conditions: the chemical reactions occur more rapidly in the humid tropics and subtropics than in cold and arid climates.

For engineering purposes, weathered rocks are classified into different groups as shown in Table 3.2.

As weathered rocks pose a threat to the stability of engineering structures, it is important to carefully map the weathering profiles during site exploration. An example of the weathering profile that consists of a few distinct zones is given in Figure 3.10.

Table 3.2 Description of weathering degrees (after ISRM, 1981)

Term	*Description*
Fresh (F)	No visible sign of weathering.
Slightly weathered (SW)	Rock may be discolored and may be somewhat weaker than in its fresh conditions.
Moderately weathered (MW)	Less than half of the rock material is decomposed and/or disintegrated to a soil.
Highly weathered (HW)	More than half of the rock material is decomposed and/or disintegrated to a soil.
Completely weathered (CW)	All rock material is decomposed and/or disintegrated to a soil.

Figure 3.10 Weathering profile showing different zones of rock weathering: CW – completely weathered, HW – highly weathered, MW – moderately weathered, and SW – slightly weathered. For zone description, refer to Table 3.2.

Question: *Is there any type of rock that is very susceptible to weathering?*
Answer: Volcanic rocks, especially granite, tend to develop deep weathering profiles (20–30 m deep). The reason for this is that the mineral composition of such rocks includes elements which are rather vulnerable to chemical weathering. We will discuss the mineral composition of such rocks in the following chapters.

3.4 Project work: geological structures and rock weathering

Geological structures can be identified during site investigation by the examination of geological features of the studied area. Faults and folds can be detected from the analysis of geological maps of the area. This will be discussed in Chapter 5.

Examination of borehole logs (Figures 2.2–2.5) reveals that the top layer consists of heavily (highly) weathered sandstone that may cause stability issues during construction. From the borehole description of the sandstone layer, it can be inferred that this layer is stratified; that is, the bedding planes of sandstone alternate with thin layers of shale (flysch). This may be one of the reasons why these geological strata are heavily weathered, as water can relatively freely enter the rock mass through the bedding joints. In addition, different responses of sandstone and shale to weathering may generate additional stresses, resulting in higher levels of rock disturbance.

3.5 Review quiz

1. Earthquakes are caused by the

   a) movement of the plates b) movement of the asthenosphere
   c) movement of the core d) all the processes above (a–c)

2. The upper mantle is

   a) solid b) liquid c) gas d) none of a–c

3. The two weathering processes are

   a) water and air b) physical and chemical
   c) streams and rivers d) nature and human

4. Which statement is NOT correct?

   a) The exposed rocks at the surface of the earth are subject to weathering
   b) Weathering causes rocks to become more porous
   c) Volcanic rocks are more resistant to weathering than other types of rock
   d) Soils are the result of rock weathering

5. Bending of strata due to regional subsidence or uplift produces:

   a) folds b) joints c) faults d) all three (a–c)

6. Which statement is NOT correct?

   a) A bedding plane is a discontinuity separating mostly igneous rocks
   b) Bedding planes can be closed and cemented
   c) Bedding planes represent interruption in the course of rock formation
   d) Bedding planes could become potential weathered zones and pockets of ground water

7. Which rocks often have beddings?

   a) basalt b) granite c) tuff d) sandstone

Answers: 1. a 2. a 3. b 4. c 5. d 6. a 7. d

Chapter 4

Rocks and rock minerals

Project relevance: Examination of the borehole logs indicates that the rock mass consists of three rock types such as sandstone, mudstone and andesite. What are these rocks? What is their origin? To find answers to these questions, it is important to become familiar with common rock types and be able to identify them during site exploration. This chapter will explain how rocks are formed and discuss different methods of identifying the most common rocks and minerals.

4.1 Rock minerals

All rocks are composed of minerals, which are naturally occurring inorganic substances with a definite composition. Rock-forming minerals are mostly silicates, while the remainder are carbonates, oxides, hydroxides and sulfates. Although some minerals are found in rocks in very small fragments (less than 1 cm), they still play a very important role in the rock behavior. The Bergeforsen Dam can serve as an example of special problems caused by mineralogy, where basic rocks under the dam decomposed in a few years after impoundment of water due to rapid decay of calcite. As a result, the originally hard rock transformed to clay (Goodman, 1989), partially lost its strength and was no longer able to sustain the load from the engineering structure.

Question: *How many rocks and minerals should be known?*
Answer: Although there are more than 200 minerals and 1000 types of rock, for civil engineering purposes it is sufficient to know the most common rock-forming minerals (about 15–20) and rocks (about 40). You should become familiar with the properties of these minerals and rocks and to be able to identify them in practice.

Question: *What is the most common rock-forming mineral?*
Answer: Quartz and feldspar. Quartz is one of the most common forms of silica. Unlike many rock-forming minerals, it is resistant to weathering and can be seen in both rock and soil. High quartz content in rock indicates high strength and hardness.

There are some distinctive features of minerals such as color, streak, luster, hardness and cleavage that are used for identification.

Color. Although the color can be helpful for certain minerals (e.g., pyrite, which has a color close to that of gold), it is generally considered to be the least reliable diagnostic property of minerals because some minerals can have different colors. For example, calcite can be white, pink or blue.

Streak is the color of a mineral's powder, which can actually be different from the mineral's color. For example, for a dark hematite, the streak is red-brown, while for variations of calcite (which can be of different color), the streak is white.

Question: *How useful is the streak in identification of minerals?*
Answer: Streak can be helpful to identify a few minerals. A good example is pyrite, which is also known as 'fool's gold'. Pyrite looks like gold (that is why people tend to assume that it is gold); however, it has a black streak unlike the gold mineral with a yellow streak.

Luster describes the way light is reflected off a mineral's surface. It can be metallic, vitreous (glassy), greasy, silky or dull.

Hardness of a mineral is its resistance to being scratched. Mohs' scale of hardness (Table 4.1) varies from 1 (talc – the weakest) to diamond (10 – the hardest). A hardness pick set that consists of picks with every level of hardness is regularly used to identify the mineral hardness.

Question: *What should you do if you don't have a hardness pick set?*
Answer: It is possible to assess mineral hardness using other methods: your fingernail (hardness of about 2.5), coins (3.5), knife blade or nail (5.5) or glass (6).

Cleavage is the way that minerals break along well-defined planes of weakness. There are different ways to describe cleavage (Figure 4.1): (a) quartz lacks cleavage (rough or uneven surface on rupture); (b) mica has one direction of easy cleavage given by its sheet structure; (c) calcite has three sets of cleavage directions.

Question: *Quartz and calcite seem to look alike (Figures 4.1a,c). How can we tell the difference?*
Answer: There are two features that distinguish these two minerals: (1) calcite can be scratched by a knife (and also by quartz), while quartz is harder than the knife; (2) quartz

Table 4.1 Mohs' scale of hardness

Mineral	*Hardness (H)*	*Special notes*
Talc	1	Softest mineral, can be scratched by fingernail
Gypsum	2	Can be scratched by fingernail
Calcite	3	Can be scratched by a copper penny
Fluorite	4	Can be scratched by a steel point
Apatite	5	Can be scratched by a knife or window glass, which is rated as H = 5.5 on the hardness scale
Orthoclase (feldspar)	6	Can be scratched by a knife blade
Quartz	7	Quartz scratches steel, glass and all minerals with H < 7
Topaz	8	A valuable jewelry stone
Corundum	9	Has many gem varieties including ruby and sapphire
Diamond	10	The hardest mineral known

Figure 4.1 Cleavage of common minerals: (a) quartz – no cleavage (rough, irregular surface), (b) mica – perfect cleavage in one direction; (c) calcite – cleavage in three directions

Table 4.2 Simplified mineral identification flow chart (modified after Goodman, 1989). H – hardness.

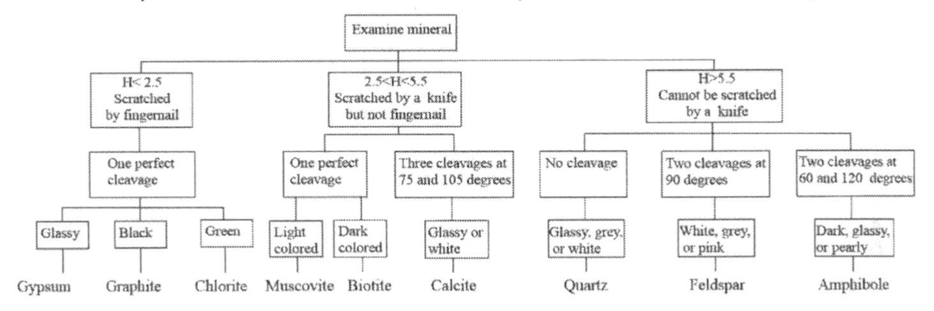

has no cleavage (uneven surface) while calcite has well-defined cleavage (it often presents rhombohedral angles between the cleavage surfaces).

4.2 Identification of common rock-forming minerals

Considering the aforementioned mineral features, it is possible to identify common rock minerals during site investigation. Hardness (H) and cleavage can provide important information about the mineral type. The simplified chart (Table 4.2) presents a systematic approach that can be used for mineral identification.

Question: *Are the rock-forming minerals from this simplified chart also present in soil?*
Answer: Yes, but not all of them. Unlike quartz, which is resistant to weathering, most rock-forming minerals transfer to secondary (clay) minerals over time due to chemical weathering. A common example includes feldspars that alter into kaolinite (clay mineral).

4.3 Rock cycle and rock types

Rocks are formed in three different ways: (1) igneous rocks are made from magma, (2) sedimentary rocks from sediments, and (3) metamorphic rocks through metamorphism (Figure 4.2). The rock cycle involves the following processes: when magma (molten rock) approaches the surface, volcanic eruptions occur, forming volcanic (extrusive) rocks (Figure 4.2a). If magma stops short, slow cooling in place produces intrusive (or plutonic) rocks. Due to weathering, igneous rocks disintegrate to soil (small-sized clasts) and are washed off to the sea/ocean by streams or rivers (Figure 4.2b). These sediments gradually accumulate at

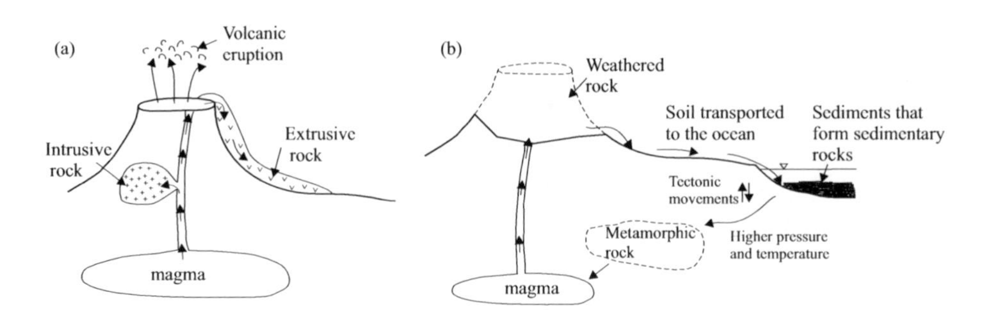

Figure 4.2 Schematic illustration of rock formation: (a) formation of igneous rocks, (b) rock cycle

the bottom of the ocean over geological time, densify under high pressures and temperatures and form sedimentary rocks. Due to tectonic processes, the sedimentary rocks can be subjected to much higher temperatures and pressures, resulting in the formation of metamorphic rocks. The following chapters will discuss the properties of each rock type.

4.3.1 Igneous rocks

Igneous rocks are formed from magma when it cools and solidifies (Figure 4.2a). The movement of silicate magma creates two types of igneous rocks: *intrusive* (or *plutonic*) rocks formed by slow cooling at depth and *extrusive* rocks formed on the surface in volcanic eruption. *Intrusive* rocks include granite, diorite and gabbro while common *extrusive* rocks are basalt, andesite and rhyolite.

Igneous rocks are composed of a relatively small number of minerals such as feldspars, quartz, pyroxenes, hornblende, olivine, and biotite. As magma cools, igneous rocks develop an interlocking *crystalline texture* due to the growth of crystals from a melt solution. Such a structure can provide high strength, but it can also be rather susceptible to weathering.

Question: *How can we then distinguish between intrusive and extrusive types?*
Answer: By the size of crystals (Figure 4.3). Intrusive rocks (e.g., diorite, Figure 4.3a) that have cooled slowly develop coarse (large) crystals because there is sufficient time for crystals to grow. In contrast, extrusive rocks (e.g., rhyolite, Figure 4.3b) cool relatively quickly, resulting in much smaller-sized crystals.

For igneous rocks, the color can serve as an indicator of the rock type because it strongly depends on the silica content – that is, the more silica, the lighter the color is (Table 4.3).

Question: *Granite and diorite look similar in terms of color. What is the difference between these rocks?*
Answer: Granite and diorite are light-colored rocks. However, unlike granite that contains abundant quartz, diorite has very little or no quartz. In diorite, the dark mineral pyroxene occurs along with feldspar, giving the rock a 'salt-and-pepper' appearance (Figure 4.3a).

During eruption (Figure 4.2a), gas bubbles in the molten lava become voids in the solidified form, producing light and porous volcanic rocks such as *pumice and scoria* (Figure 4.4).

Figure 4.3 Different sizes of crystals found in igneous rocks: (a) coarse black-and-white crystals in the intrusive rock of *diorite*, creating a 'salt-and-pepper' look; (b) very fine-sized crystals in the extrusive rock of *rhyolite*

Table 4.3 Relationship between the rock color and silica content for major intrusive and extrusive rocks

Intrusive rock	*Extrusive rock*	*Silica content (%)*	*Color*
Granite	Rhyolite	> 65	Light
Diorite	Andesite	50–65	Combination of light and dark
Gabbro	Basalt	40–50	Dark

Figure 4.4 Porous volcanic scoria

Volcanic glass (also known as obsidian) arises from the sudden cooling of lava droplets during flight.

Question: *Is there any practical use for such light volcanic rocks?*
Answer: Yes, pumice is widely used in engineering practice to make lightweight concrete, while obsidian was used in ancient times to make sharp objects such as knives, arrowheads, spear points and other weapons and tools.

4.3.2 Sedimentary rocks

Sedimentary rocks are composed of discrete fragments (clasts) of materials derived from other rocks. These clastic sediments are converted into rock or rocklike material through the process of lithification, where the water is squeezed from the pore space by the increasing pressure of overlying sedimentation.

Sedimentary rocks generally have a *clastic texture*, which is a collection of rock and mineral fragments. It is possible to distinguish a clastic from a crystalline texture through the examination of a clean, fresh rupture surface.

Question: *What texture, crystalline or clastic, generally produces rocks with higher strength?*
Answer: Crystalline rocks with low porosity prove to be more competent and stronger rock types compared to sedimentary rocks with a clastic texture. However, crystalline rocks (e.g., granite) can be more susceptible to weathering.

Wherever erosion of existing rocks occurs, the products are carried by streams into rivers and accumulate in the oceans. These soft soil deposits eventually harden into sedimentary rock through compaction, consolidation and sedimentation. Among the sedimentary rocks, the most widespread are sandstone, shale, siltstone and mudstone.

Question: *How can we distinguish sandstones from siltstones and mudstones?*
Answer: Sandstones generally have a friable quality, meaning that grains can be dislodged by rubbing the rock surface with a thumb. Rocks made of finer clasts that cannot be felt with a thumb are likely to be siltstones or mudstones.

Question: *How do we distinguish shales from mudstones? Aren't they both argillaceous rocks?*
Answer: Yes, argillaceous rocks like shale and mudstone are made of fine particles. However, unlike mudstones, shales display the property of fissility, meaning that they tend to split easily.

Another commonly occurring sedimentary rock is limestone. It consists of calcium carbonate ($CaCO_3$), and it is commonly formed from accumulation of lime shells from shellfish.

Question: *Is dolomite different from limestone?*
Answer: Yes. Dolomite is formed when magnesium replaces part of the calcium in limestone. Dolomite rock ($CaMg(CO_3)_2$) is harder, heavier and less soluble than limestone.

Limestone can have either clastic or crystalline textures, and its color can vary from white to gray or black. As limestone is relatively soluble, solution cavities in such rocks (karst topography) may be abundant (Figure 4.5). It is noted that limestone has a wide use in construction; it is the basic ingredient in the manufacturing of cement and lime.

There are a few distinctive features of sedimentary rocks that can be used for identification:

- Bedding planes that develop in sedimentary rocks;
- Some sedimentary rocks such as limestone, rock salt or rock gypsum can be soluble in slightly acidic liquids or water;
- Some sedimentary rocks contain *fossils* (Figure 4.6). Fossils occur in many kinds of rocks but are often best preserved and most numerous in limestone, which is a sedimentary rock composed largely of the shells of animals.
- Sedimentary rocks are commonly classified by their particle size, as shown in Table 4.4.

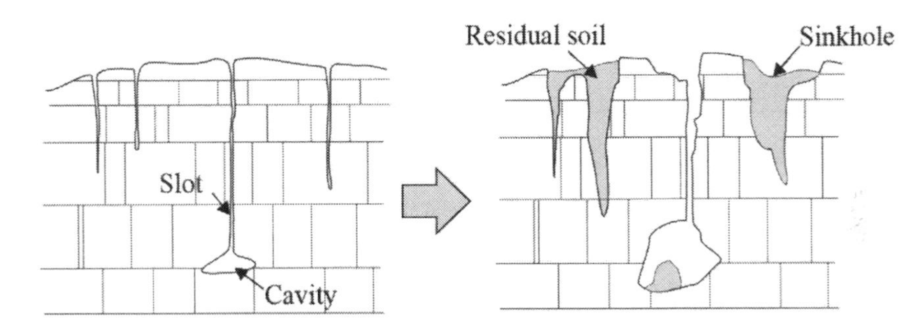

Figure 4.5 Karst development in limestone

Figure 4.6 Fossil in shale

Table 4.4 Characteristics of common sedimentary rocks

Rock name	*Particle size (mm)*	*Comments*
Conglomerate	> 2	Rounded rock fragment
Breccia		Angular rock fragment
Sandstone	0.06–2	Quartz with other minerals
Shale	< 0.06	Split into thin layers
Mudstone		Break into clumps or blocks

Question: *Sedimentary rocks such as shale and mudstone are made of fine particles (clay) which are hardened and cemented. It is known that clay tends to absorb water and swell. What will happen when these rocks come in contact with water?*
Answer: Shales, mudstones and related rocks frequently soften when soaked in water and yield under relatively low stresses. This can also cause engineering problems such as heave and lifting of light engineering structures.

Question: *Are clay minerals also found in rocks? Or are rocks only made of rock-forming minerals such as quartz, feldspars, etc.?*
Answer: Yes, clay minerals are also present in rocks. Clay minerals are produced in weathering by decomposition of feldspar and ferromagnesian minerals such as pyroxene, amphibole, olivine, and mica. These clay minerals are then carried in the clay-sized sediments of streams out into the ocean where they settle, accumulate and solidify into sedimentary rocks.

Question: *Are there any tips for telling the difference between breccia and conglomerate?*
Answer: Breccia includes mostly angular rock fragments (Figure 4.7), while conglomerates mostly contain rounded rock fragments.

4.3.3 Metamorphic rocks

When rocks are subjected to high temperatures and pressures, their texture and mineralogy may change so dramatically as to create a new type of rock: a metamorphic rock. Under high temperatures and pressures, the porosity of rock decreases, its strength increases and the unit weight of the solid material increases due to the loss of chemically bound water. In addition, many metamorphic rocks develop a strong directional structure (anisotropy), in which rock properties (strength, Young's modulus etc.) depend on direction. In such metamorphic rocks (e.g., slates and schists), new minerals are developed and arranged in a parallel order, forming foliation.

Metamorphic rocks are formed from sedimentary or igneous rocks, as shown in Table 4.5. There are different grades of metamorphic rocks, which are defined by the temperature. As the temperature during metamorphism increases, the following rock transformation occurs:

$$shale(sedimentary\ rock) \rightarrow slate\ (low\ grade) \rightarrow schist(intermediate\ grade) \rightarrow gneiss(high\ grade)$$

Figure 4.7 Breccia with angular rock fragments

Table 4.5 Characteristics of common metamorphic rocks

Metamorphic rock	*Original rock*	*Texture*	*Metamorphic grade*
Slate	Shale	Foliated	Low
Mica schist	Shale	Foliated	Low to intermediate
Chlorite schist	Basalt	Foliated	Low
Gneiss	Granite, shale	Foliated	High
Marble	Limestone, dolomite	Non-foliated	Low to high
Quartzite	Quartz sandstone	Non-foliated	Intermediate to high

In this chain, shale is the weakest rock while gneiss is the hardest rock.

Question: *Do anisotropy and foliation mean the same thing?*
Answer: Not exactly. Anisotropy is a more general term, meaning different rock properties in different directions. Foliation is a distinctive feature of metamorphic rock which is related to its structure. It is formed under high pressures and temperature when newly developed minerals are arranged in a parallel order (Figure 4.8).

Question: *Any tips on how to identify metamorphic rocks?*
Answer: Similar to igneous rocks, metamorphic rocks may have well-defined crystals; however, unlike igneous rocks, many metamorphic rocks have a foliated texture.

Figure 4.8 Bands in gneiss

4.4 Identification of common rocks

Question: *Is it expected to identify all types of rocks found during site investigation? Any tips on how to do it?*

Answer: It may be rather difficult or even impossible to correctly name all rock specimens, as sometimes it requires not only special training in petrology but also special petrographic examination of some rock specimens. You can get some idea about common rock types and geology by examining geological maps of the project area prior to site investigation. The chart in Figure 4.9 can assist you with identifying rock specimens while examining their fresh surface. The most important identification features will be texture, hardness and structure.

4.5 Engineering problems related to rocks

4.5.1 Engineering problems related to igneous rocks

Intrusive rocks are usually sufficiently strong for any engineering purpose when they are fresh. However, these rocks tend to decompose to significant depths due to weathering over geologic time. Igneous rocks change their mineral composition during weathering, forming a mixture of sand, silt and clay. Most of the rock-forming minerals, except for quartz and mica, are removed during chemical weathering, which results in a relatively high rate of weathering.

4.5.2 Engineering problems related to sedimentary rocks

- *Heave of shale and mudstone.* During construction, stripping of the surficial soil may expose expansive bedrock such as mudstone and shale to rainwater, causing heave

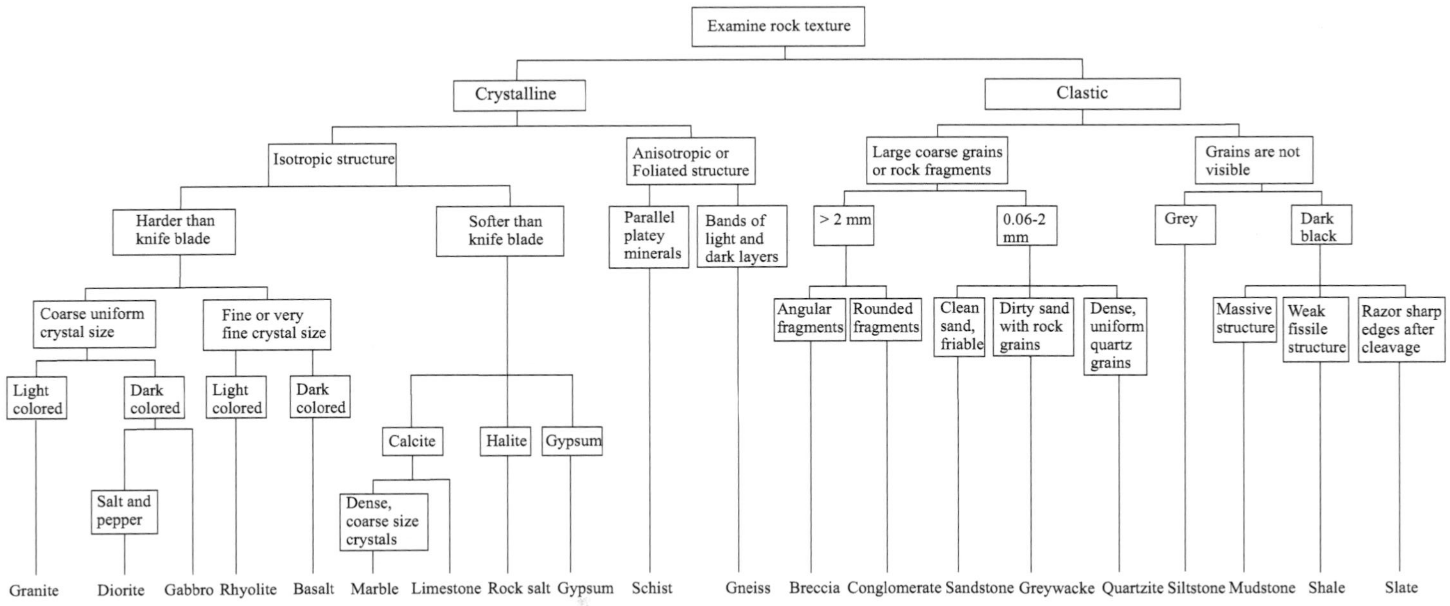

Examine rock texture

Crystalline
Clastic

Isotropic structure
Anisotropic or Foliated structure
Large coarse grains or rock fragments
Grains are not visible

Harder than knife blade
Softer than knife blade
Parallel platey minerals
Bands of light and dark layers
> 2 mm
0.06-2 mm
Grey
Dark black

Coarse uniform crystal size
Fine or very fine crystal size
Angular fragments
Rounded fragments
Clean sand, friable
Dirty sand with rock grains
Dense, uniform quartz grains
Massive structure
Weak fissile structure
Razor sharp edges after cleavage

Light colored
Dark colored
Light colored
Dark colored
Calcite
Halite
Gypsum
Salt and pepper
Dense, coarse size crystals

Granite
Diorite
Gabbro
Rhyolite
Basalt
Marble
Limestone
Rock salt
Gypsum
Schist
Gneiss
Breccia
Conglomerate
Sandstone
Greywacke
Quartzite
Siltstone
Mudstone
Shale
Slate

Figure 4.9 Simplified chart to identify commonly occurring rocks

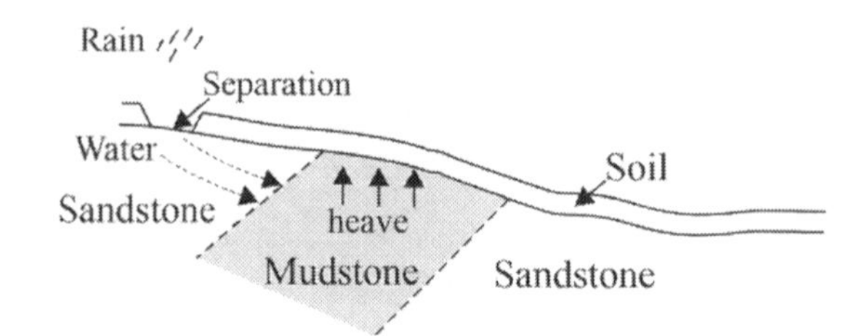

Figure 4.10 Schematic illustration of heave in interbedded sandstone and mudstone (after Meehan et al., 1975)

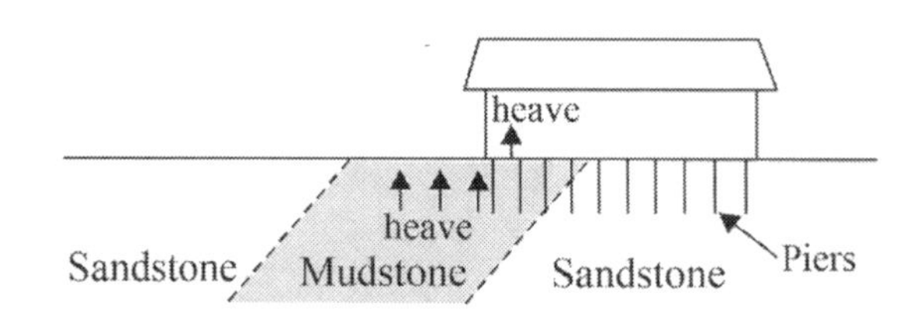

Figure 4.11 Damage to house on shallow piers (after Meehan et al., 1975)

(Figure 4.10). This can result in cracked streets, cracked house foundations and broken driveways and walks. Although mudstone is considered impervious, it can be jointed and/or interbedded with pervious sandstone which can conduct water.

Question: *Is it possible to avoid or minimize heave?*
Answer: It is important to provide and maintain drainage. Engineering solutions may also involve foundations supported by reinforced piers (Meehan et al., 1975), as schematically shown in Figure 4.11.

- *Slope stability issues.* Excavation in shales conducted in Bogotá (Colombia) revealed that the shales did not swell and heave but rather quickly deteriorated into fat clay (very plastic and soft soil), which resulted in increased landslide hazard along roads and pipelines (Goodman, 1993).
- *Squeezing of soft sedimentary rocks.* Squeeze refers to the gradual reduction in the cross-sectional area of tunnels, a process that is commonly associated with shales and mudstones. The tunnel section may be so reduced as to require additional support. If a tunnel machine is used, it may become stuck.
- *Slaking of argillaceous rocks.* Shales and mudstones can deteriorate and break after exposure to excavation; that is, they can slake. Slaking can begin immediately with visible cracking and heaving. Slaking is more pronounced in shales where the fissility may open up like the pages of a book.

4.5.3 Engineering problems related to metamorphic rocks

Metamorphic rocks, when fresh, are generally sufficiently sound for almost all engineering purposes. However, some highly schistose rocks are inherently troublesome even when fresh (Goodman, 1993). Foliated rocks can be affected by landslide movements, which can be creep, topples or rock slab slides. This process is controlled by the orientation of schistosity and can occur during excavation.

4.6 Project work: analysis of rock types

There are three types of rock at the project site: sandstone, mudstone and andesite. The former two are sedimentary rocks with a clastic texture while the andesite is an intrusive, igneous rock. The sandstone will likely have high porosity because this rock is relatively soft and weathered. The mudstone seems to be of good quality, according to the description provided in the borehole logs (Figures 2.2–2.5). It is only slightly weathered, and it has relatively high values of strength. However, it may experience some heave issues if it comes in contact with water. The fresh andesite is of high strength, and it may not cause engineering problems during construction.

4.7 Review quiz

1. Igneous rocks are formed from

   a) pressure b) sediments c) magma d) all three (a–c)

2. According to the rock cycle, there are three types of rocks (by origin), which are called sedimentary, metamorphic and

   a) plutonic b) intrusive c) igneous d) foliated

3. According to the Mohs' scale of hardness, the weakest mineral is

   a) gypsum b) talc c) calcite d) gold

4. This mineral is commonly known as 'fool's gold' because of its similarity in color and shape with gold. However, it has a black streak.

   a) silver b) sulfur c) pyrite d) selenite

5. It is the most common mineral found on the surface of the Earth. It is the main ingredient in sand, sandstone and granite.

   a) calcite b) quartz c) mica d) feldspar

6. The clastic texture can be found in:

   a) basalt b) sandstone c) granite d) andesite

7. When sea animals die, their shells and bones form layers of sediment on the ocean floor. This buildup of sediments forms

   a) siltstone b) conglomerate c) mudstone d) limestone

8. Which one of the following igneous rocks does not belong to the intrusive type?

   a) granite   b) diorite   c) basalt   d) gabbro

9. Which of the following is not metamorphic rock?

   a) slate   b) marble   c) gneiss   d) rhyolite

10. It is a smooth, glassy extrusive rock that cooled very quickly, leaving no time for crystals to grow. Early people used it for tools and arrowheads because of its sharp edges. What is it?

    a) tuff   b) pumice   c) obsidian   d) scoria

11. It is an extremely hard, medium to coarse-grained intrusive rock. It is composed of light-colored and dark-colored crystals which give it a 'salt-and-pepper' effect. What is it?

    a) diorite   b) granite   c) andesite   d) basalt

12. It is the most common extrusive rock that forms on the surface of the Earth. It is formed from lava that pours out of volcanoes and cools quickly. It is dark, and it contains only very small crystals. What is it?

    a) basalt   b) granite   c) andesite   d) rhyolite

13. What rock is considered soluble?

    a) gypsum   b) siltstone   c) mudstone   d) argillite

Answers: 1. c 2. c 3. b 4. c 5. b 6. b 7. d 8. c 9. d 10. c 11. a 12.a 13. a

Chapter 5

Rock exploration

Project relevance: Site investigation is an essential part of the project work as it generates important data about geological structures and rock mass properties. This chapter will introduce common methods of site investigation and discuss engineering issues related to different types of rock mass. The project data from borehole logs will be analyzed and a subsurface profile of the project site will be created.

5.1 General considerations

Site investigation is conducted to assess the surface and subsurface conditions at proposed construction sites, especially where the geologic conditions may vary greatly. Some sites may present geologic conditions where soil cover is either thin or absent and where observations of outcrops reveal all the rock types and structures. On the other hand, sites in the tropics may have a thick, residual soil cover that makes it rather difficult to study the underlying rock formations. In any case, field observations of outcrops, deep drilling and geophysical surveys will provide necessary information to create a site geologic model.

Question: *How much investigation is required to create a reliable geologic model?*
Answer: The amount of investigation depends on several factors such as project budget, time frame and type of engineering structures. With sufficient time and money, it is possible to create a very detailed model; however, the funding is never limitless, and the project time is usually constrained. The scope of investigation depends on the type of engineering structures and expected geologic conditions. Nevertheless, no matter what is to be built, there are common targets of exploration that include the stratigraphy and geological structures, the type of rocks and the location of boundaries between them, the thickness and character of weathered material, and the level of ground water.

5.2 Desk study

Desk study is the first step of investigation when all relevant information, including aerial photos and maps, is collected and analyzed. The aerial photos of the investigated site taken in different years can provide important data about geologic faults as well as natural processes such as sinkholes, landslides and erosion. Geologic maps are essential in creating

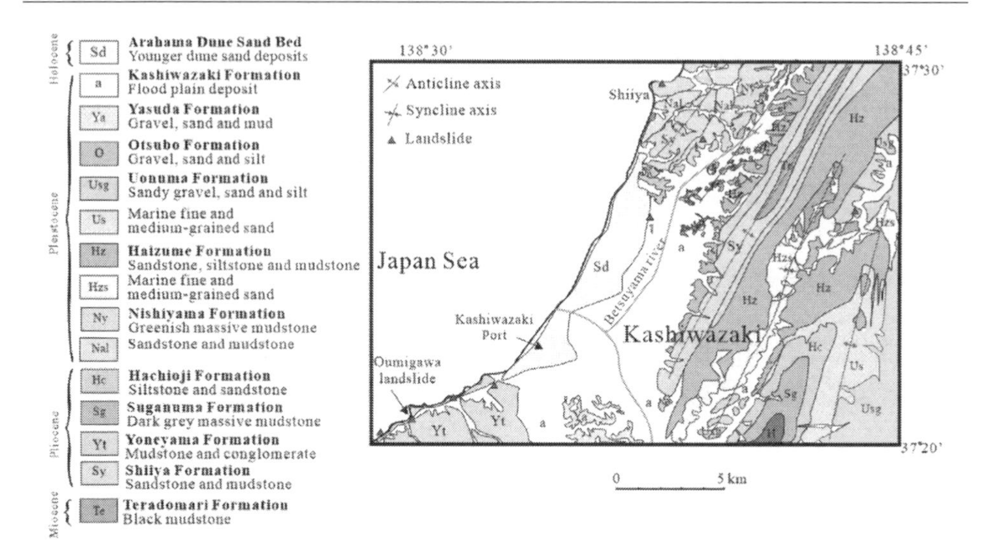

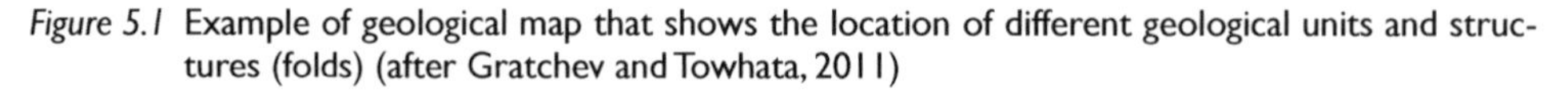
Figure 5.1 Example of geological map that shows the location of different geological units and structures (folds) (after Gratchev and Towhata, 2011)

a geological model of the project site; they should be available at the domestic Geologic Survey Office. Such maps indicate the type of rocks and boundaries between bedrock units (Figure 5.1). They also include cross-sections of the area that provide interpolation of geologic structures and stratigraphy.

Question: *Are there any guidelines for a geological map scale?*
Answer: Yes, geological maps are prepared for various purposes and thus they have different scales. Maps with a scale of 1:100,000 are regarded as synoptic maps as they cover larger areas. Maps with scales of 1:20,000 and 1:50,000 are basic maps, while detailed maps with scales from 1:2,000 to 1:10,000 may cover only the project area.

5.3 Field work

Field investigations are carried out in accordance with the relevant standard that clearly outlines the field procedures. Field work typically includes boreholes, site testing and seismic exploration.

5.3.1 Boreholes

Boreholes are an important part of site investigation as they provide information about the geological settings of the area. The number of boreholes necessary to create a reliable geological model tends to increase when (a) the rock/soil variability increases, (b) the loads from engineering structures increase, and (c) very important structures (e.g., dams) need to be designed.

Question: *How close to each other and how deep should boreholes be?*
Answer: The spacing and depth of boreholes depend on site conditions and the type of engineering structure which will be constructed. The guidelines and borehole characteristics

can be found in the relative standards. Figure 5.2 schematically illustrates what may happen when boreholes are not sufficiently spaced and deep enough. This may lead to an incorrect geological model as shown in Figure 5.2a. The three boreholes spaced relatively far from each other may give the wrong impression that rock/soil layers are running almost parallel to the ground surface. However, a more detailed investigation of the same area that includes additional and deeper boreholes (Figure 5.2b) will detect bedding planes dipping at a certain angle to the ground surface, providing a more accurate geological model.

Core samples collected from boreholes provide valuable information about the subsurface geology. During an investigation, the core is laid out in order in sturdy wooden boxes and photographed. It is carefully examined using a hand lens, hammer or knife, while data on rock type, strength, degree of weathering, water and type of discontinuities is recorded in borehole logs. For civil engineering purposes, it is also important to record the character and location of zones with fractured and crushed rocks as well as the weak clay seams.

Borehole logs also provide information on the rock core quality using the RQD (rock quality designation) index proposed by Deere (1964). RQD indicates the degree of rock fracturing and it is defined as the percentage of the length drilled that yields a core in pieces longer than 100 mm (Figure 5.3). It is noted that many rock mass classification systems (see Chapter 9) use the RQD as one of their key parameters. The RQD is computed from each core run (Equation 5.1) and plotted in a strip alongside the lithological core (Figures 2.2–2.5).

$$RQD = \frac{\sum Length\ of\ core > 10\ cm\ length}{Total\ length\ of\ core\ run} \cdot 100\% \tag{5.1}$$

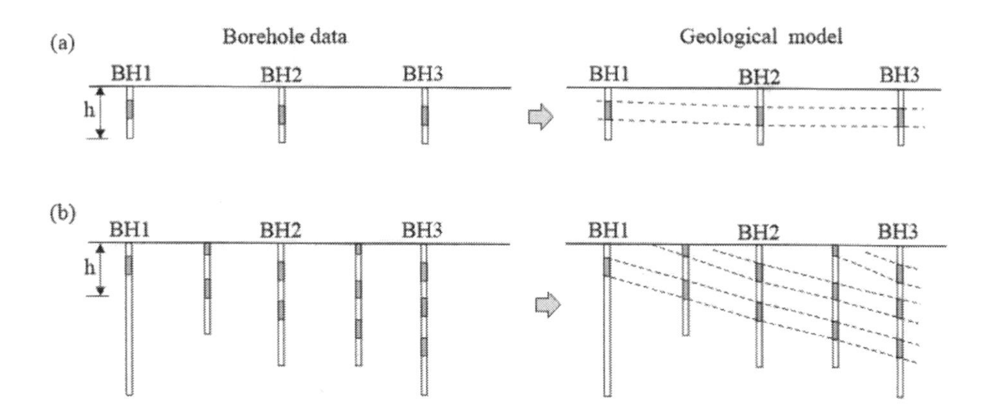

Figure 5.2 The effect of borehole characteristics on the development of a geological model

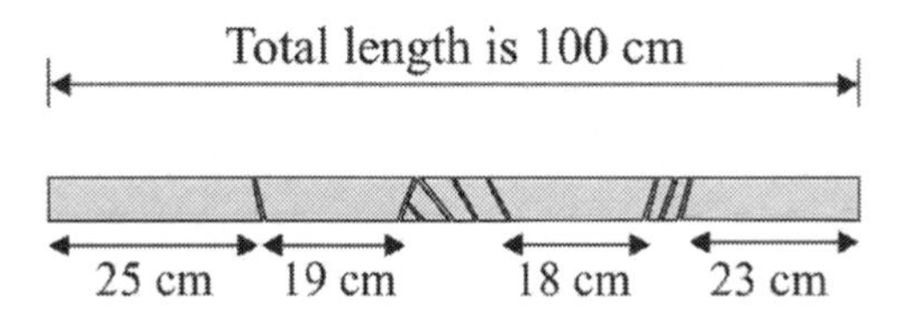

Figure 5.3 Determination of RQD using the length of core pieces

For the core run in Figure 5.3, RQD equals

$$RQD = \frac{25+19+18+23}{100} \cdot 100\% = 85\%$$

Rock mass can be classified on the basis of RQD, as shown in Table 5.1.

Simple tests can be performed on site to determine physical properties of rocks. They include moisture content tests as well as strength tests using a portable point load press (more information about point load tests is given in Chapter 7). Visual observations of the core are performed to determine the level of weathering, following the description given in Table 3.2. Table 5.2 describes common techniques used in the field to determine the strength characteristics of the core.

Table 5.1 Classification of rock core based on RQD (after Peck et al., 1974)

RQD (%)	*Rock quality*	*Allowable bearing pressure (MPa)*
0–25	Very poor	1–3
25–50	Poor	3–6.5
50–75	Fair	6.5–12
75–90	Good	12–20
90–100	Excellent	20–30

Table 5.2 Classification of rock core based on its strength

Term/Symbol	*Description*	*Point load index (MPa)*
Extremely low/EL	Easily remolded by hand to a material with soil properties	≤ 0.03
Very low/VL	Material crumbles under firm blows with sharp end of pick; it can be peeled with a knife. Pieces up to 3 cm thick can be broken by finger pressure.	$0.03 \leq 0.1$
Low/L	Easily score with a knife. A piece of core 150 mm long and 50 mm diameter may be broken by hand.	$0.1 \leq 0.3$
Medium/M	Readily scored with a knife. A piece of core 150 mm long and 50 mm diameter can be broken by hand with difficulty.	$0.3 \leq 1$
High/H	A piece of core 150 mm long and 50 mm diameter cannot be broken by hand but can be broken by a pick with a single firm blow; rock rings under hammer.	$1 \leq 3$
Very high/VH	Hand specimen breaks with pick after more than one blow; rock rings under hammer.	$3 \leq 10$
Extremely high/EH	Specimen requires many blows with geological pick to break through intact material; rock rings under hammer.	> 10

5.3.2 Seismic methods

Rock exploration can be done relatively quickly and inexpensively by means of seismic methods. It is based on the phenomenon that elastic waves generated by an explosion or by a weight incidence propagate at different velocities depending on the elastic properties and densities of rocks (Figure 5.4a). Thus, from the travel time of seismic waves, which are measured at the ground surface, the thickness of each rock strata can be determined (Figure 5.4b).

Question: *What is the practical importance of seismic methods? Why can't we only use boreholes?*
Answer: First of all, borehole exploration is the most expensive part of site investigation, while seismic methods can provide an inexpensive alternative. Also, a borehole provides data only for the point where it was drilled, which is usually not sufficient to create a reliable geological model of the whole site. Seismic methods can cover large areas and clarify the geological settings across the project site.

Question: *Is it possible to determine the type of rock using seismic methods?*
Answer: Unfortunately, no. This is one of the disadvantages of seismic methods. For this reason, a geophysical survey should be made in the area that has already been studied by a net of borings or pits so that the data from boreholes and seismic investigation can be combined and analyzed together.

The seismic method is also used to clarify profiles of weathered rocks which can be difficult to detect by borings. Weathered rocks manifest themselves by extremely low velocities of seismic waves, while hard and fresh rocks are associated with much greater values of seismic velocity (Table 5.3).

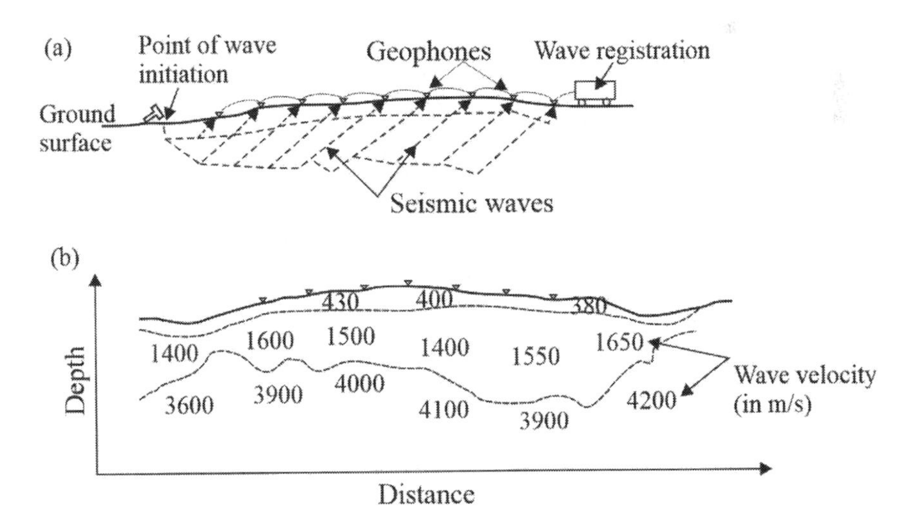

Figure 5.4 Schematic illustration of seismic test setup (a) and data analysis (b)

Table 5.3 Relationship between seismic velocity and rock strength

Seismic velocity (m/s)	*Geotechnical classification*	*Estimated unconfined compressive strength (MPa)*
< 2000	Low strength rock	< 10
2000–2500	Medium strength rock	10–20
2500–3500	High strength rock	20–60
3500–7000	Very high strength rock	> 60

5.4 Engineering issues during site investigation

From engineering practice, we know what problems can occur in certain types of rock, so that we need to keep it in mind during site investigation. Table 5.4 summarizes the most common engineering issues by rock type.

5.5 Project work: cross-section and geological units

Cross-sections provide important data on the geological conditions existing at the project site, including the type of rock, boundaries between different units and ground water. A cross-section through the A-A′ line from Figure 2.1 is given in Figure 5.5. The subsurface conditions at the project site consist of a sandstone layer (about 3–4 m thick) overlying a stratum of mudstone (whose thickness varies from 7 m to 8 m) and very hard andesite (it is not clear how thick this layer is).

The sandstone seems to be heavily fractured as its RQD value is relatively low (an average of 26%). As the sandstone is distinctly to highly weathered, its strength is estimated as low (Figures 2.2–2.5). The mudstone and andesite rock masses have much higher average RQD values of 77.5% and 92%, respectively. This implies that the rock quality of mudstone and andesite is much higher than that of the sandstone.

No data on the ground water conditions is given in the borehole logs; however, it is noted that the sandstone is moist. As we examine all four boreholes, we notice that the thickness of each layer seems to remain almost the same across the slope. We will use this observation later when we create a computer model for slope stability analysis (Chapter 11).

Question: *Do we need to connect each layer across the slope using the data from the boreholes as reference?*
Answer: It is not necessary because we do not have enough information about the boundaries between the geological layers across the whole slope.

5.6 Review quiz

1. The first phase of geotechnical investigation is to

    a) collect all available information
    b) make a site visit
    c) prepare borehole locations
    d) all of the above (a–c) need to be done simultaneously

Table 5.4 Typical engineering problems related to rocks

Rocks or rock formations	*General engineering issues*	*Engineering issues related to*		
		Slope stability	*Dams*	*Tunnels*
Sandstones	Engineers need to determine the porosity of these rocks			Rocks with high quartz content create a special health problem for miners. Silica dust is toxic, causing severe respiratory illness
Argillaceous rocks (shale and/or mudstone)	Engineers need to determine the ability of this rock to slake as well as the presence of clay minerals	Landslides commonly occur in these rocks when they are weathered and exposed to water	Dams should not be placed on these rocks, as it is almost impossible to control uplift pressures from seepage	– Squeezing of ground and deformation of tunnel support may occur in these rocks – Heave during excavation can be a big problem. Heave is driven by unloading and the upward expansion in the presence of water – Slaking makes it difficult to use some forms of support, such as rock bolt bearing plates or shotcrete
Flysch (sandstone and shale/ mudstone)	– This formation typically has an uneven weathering profile with greater depths of weathering in argillaceous rocks – The contact between these two rocks is the loci of shearing and fracturing, as the rocks have different capacities to deform – Water is confined in sandstones while shale is impermeable, resulting in the internal erosion of this rock formation	Landslides may occur in sites with sandstone and shale, where blocks of the harder sandstone slide intact on the shale		When both rocks appear in the face, the tunnel boring machine vibrates badly and has to be pulled out so that the section can be mined by hand
Volcanic rocks (granite)	The main difficulty in exploration of these decomposed rocks is characterizing their weathering profiles and properties	These weathered rocks are commonly associated with debris flow during heavy rains	Concrete dams should only be placed on sound rocks of low levels of weathering (fresh or slightly weathered)	In general, underground excavations and tunnels in these rocks are constructed successfully
Metamorphic rocks (schist or gneiss)	– In these rocks, persistent shear zones typically occur parallel to the plane of foliation – Weathering typically opens the rock along the foliation	Creep is common for strongly foliated rocks		

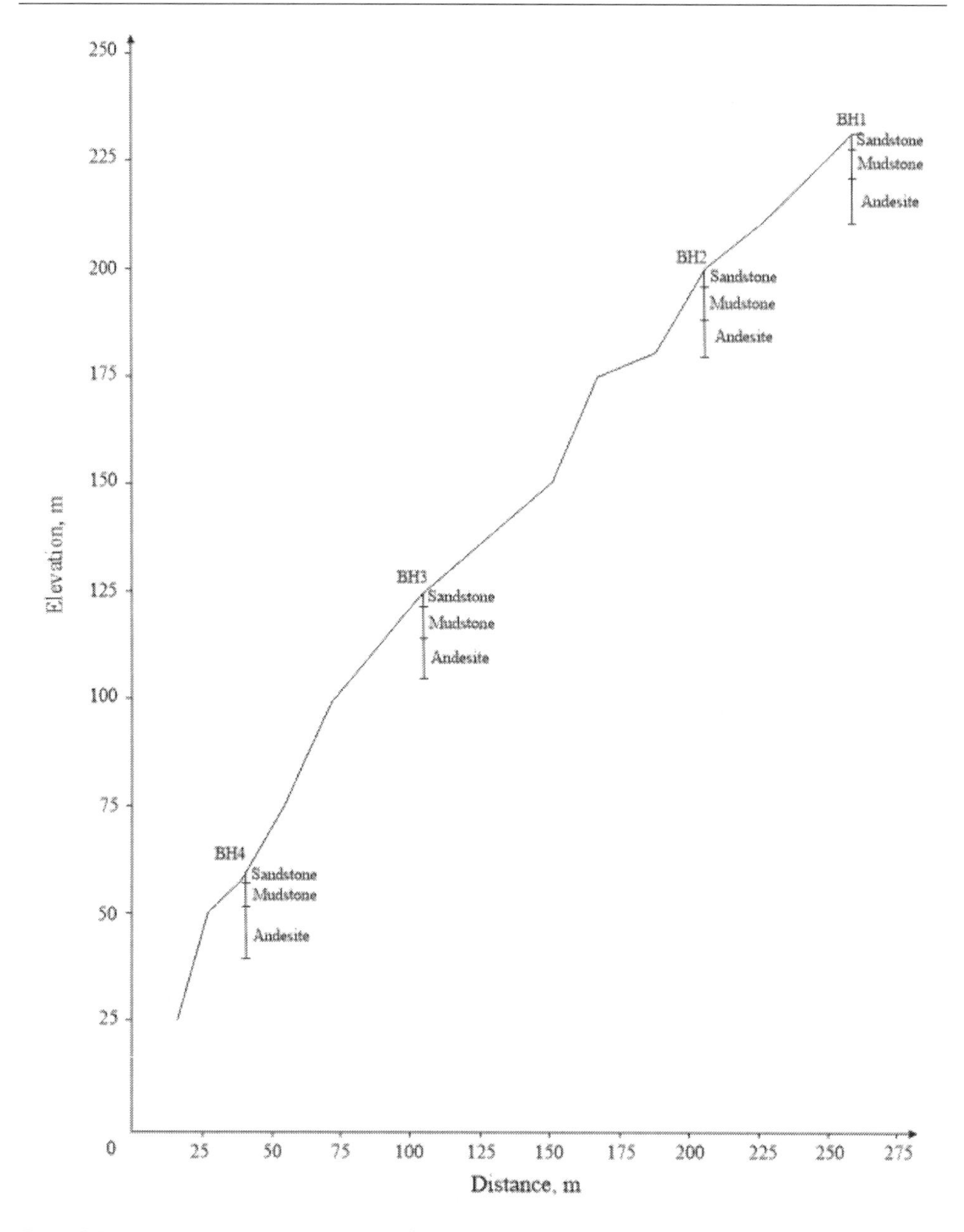

Figure 5.5 A cross-section through the A-A′ line in Figure 2.1

2. What landforms can be recognized from the interpretation of aerial photos?

   a) landslides b) faults
   c) sinkholes d) all of these (a–c)

3. The larger the RQD, the larger the joint spacing.

   a) True b) False

4. This rock with high quartz content creates a special health problem for miners. Silica dust is toxic, causing severe respiratory illness.

   a) sandstone b) mudstone c) basalt d) limestone

5. What rock will be the most susceptible to slaking during excavation?

   a) shale b) sandstone c) granite d) gneiss

6. What level of rock strength is described as "A piece of core 150 mm long by 50 mm diameter cannot be broken by hand but can be broken with pick with a single firm blow"?

   a) VH b) M c) EH d) H

7. When describing the coating or infilling of joints, what term is used to match the following description "a visible coating or infilling of soil or mineral substance but usually unable to be measured (less than 1 mm)."

   a) Clean b) Stain c) Veneer d) Coating

8. In these rocks, heave can commonly occur during excavation due to unloading and the upward expansion in the presence of water.

   a) shale b) sandstone c) granite d) gneiss

9. In these rocks, weathering typically opens the rock along the foliation.

   a) sandstone b) mudstone c) granite d) schist

10. Which statement about site investigation using geophysics is NOT correct?

    a) Geophysics can determine the type of rock
    b) Geophysics can estimate the strength of rocks
    c) Geophysics can determine the boundaries between different geological units

Answers: 1. a 2. d 3. a 4. a 5. a 6. d 7. c 8. a 9. d 10. a

Chapter 6

Discontinuities in rock mass

Project relevance: Rock is hard material that can withstand high stresses. However, rock mass always has discontinuities which decrease the overall strength of rock. It is of high importance to record the type of discontinuities and their characteristics during site investigation and determine their effect on rock mass properties. This chapter will discuss the origin of discontinuities and explain how to identify and measure them in the field. The effect of discontinuities on the rock mass properties of sandstone, mudstone and andesite at the project site will also be discussed.

6.1 Types of discontinuities

Almost every rock has mechanical defects, which generally consist of closely spaced discontinuities which are known as joints. In rock mass, joints typically exist in sets (Figure 6.1). In rock mass with innate mechanical defects such as bedding or cleavage planes, the joints constitute a source of weakness and can be the source of instability.

Question: *From an engineering point of view, what effect can joints have on rock mass?*
Answer: Joints break the continuity of the rock and reduce the average strength of the jointed mass to a small fraction of the same rock in an intake state (Gratchev et al., 2016).

Question: *Does the term joint also mean shear plane?*
Answer: No, the term joint indicates a fracture along in which no noticeable displacement has occurred.

Joints can be formed due to (a) erosion of exposed rock surfaces, (b) removal of overlying rocks (e.g., a decrease in the overburden stress), and (c) cooling of hot rock masses. In igneous rocks, which cooled rapidly, the joints are closely spaced. A good example is columnar basalt, which consists of columns oriented at right angles to the surface of cooling. Sedimentary rocks commonly contain three sets of joints – one of each is parallel to the bedding planes while the others commonly intersect the planes at approximate right angles.

Figure 6.1 Joint sets (i.e., a group of parallel joints) in rock mass

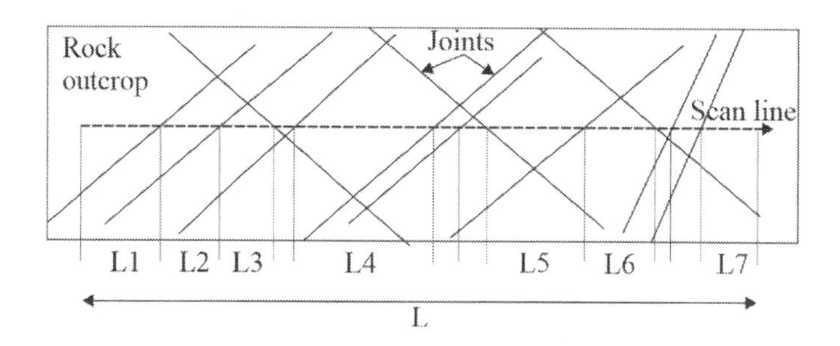

Figure 6.2 A schematic diagram showing how to obtain RQD during a scanline survey. Refer to Equation 6.1

There are two common approaches used to investigate the joint characteristics of rock outcrops: cell mapping and scanline mapping. The former is related to mapping of fracture-set properties observed within user-defined cells on the rock exposure. The latter involves the detailed mapping of individual discontinuities that intersect a designated mapping line. When surveying a rock face, it may be possible to obtain the RQD of the rock mass by measuring the spacing between the joints (Figure 6.2 and Equation 6.1).

$$RQD = \frac{L1 + L2 + L3 + L4 + L5 + L6 + L7}{L} \cdot 100\% \tag{6.1}$$

where $L1$, $L2$, . . . $L7$ are equal to or greater than 10 cm.

6.2 Joint characteristics

The important joint characteristics are spacing, persistence, joint roughness, aperture and joint orientation.

6.2.1 *Joint spacing and frequency*

Joint spacing (Figure 6.3) is defined as the perpendicular distance between the two neighboring joints. It is measured by a ruler and it can vary from a few centimeters to meters.

Classification of joint spacing commonly used in practice is given in Table 6.1.

Figure 6.3 Measurement of joint spacing in the field. Joint spacing for a pair of joints is the perpendicular distance between the two joints.

Table 6.1 Classification based on the spacing between discontinuities

Description	*Joint spacing (m)*
Extremely close spacing	< 0.02
Very close spacing	0.02–0.06
Close spacing	0.06–0.2
Moderate spacing	0.2–0.6
Wide spacing	0.6–2
Very wide spacing	2–6
Extremely wide spacing	> 6

Joint frequency (λ) is defined as the number of joints per meter length. It is the inverse of joint spacing (s_j), as given in Equation 6.2.

$$\lambda = \frac{1}{s_j} \tag{6.2}$$

Hudson and Priest (1979) proposed the following correlation between RQD and joint frequency (Equation 6.3):

$$RQD = 100e^{-0.1\lambda}(0.1\lambda + 1) \tag{6.3}$$

6.2.2 Joint persistence

Persistence is the length of a joint that can be quantified by observing the trace lengths of discontinuities on exposed surfaces. Persistence is an important parameter as it controls large-scale sliding or failure of slopes. The classification based on joint persistence is given in Table 6.2.

6.2.3 Volumetric joint count

Block size in rock mass depends on the number of discontinuity sets that separate the blocks, spacing and persistence. It can affect the rock mass stability and it is a key parameter in rock mass classification. Block size can be classified by the volumetric joint count (J_v), which is defined as the number of joints per cubic meter of rock mass (Table 6.3).

Table 6.2 Joint description based on its persistence

Description	*Trace length (m)*
Very low persistence	< 1
Low persistence	1–3
Medium persistence	3–10
High persistence	10–20
Very high persistence	> 20

Table 6.3 Block sizes and J_v values (Barton, 1978)

J_v *(joints/m³)*	*Description*
< 1	Very large blocks
1–3	Large blocks
3–10	Medium-sized blocks
10–30	Small blocks
30–60	Very small blocks
> 60	Crushed rock

Palmström (1982) suggested that when no core is available, the RQD of rock mass can be estimated from the number of discontinuities per unit volume which are visible in surface exposures. The suggested relationship is shown in Equation 6.4.

$$\mathrm{RQD} = 115 - 3.3J_v \tag{6.4}$$

Palmström (2005) provides some correlations between block volume, joint spacing and RQD which can be useful in practice to estimate the joint characteristics (Figure 6.4).

6.2.4 Aperture

Aperture is the open space between two joints (Figure 6.5). Open or filled (Figure 6.6) joints with large apertures have low shear strength.

Classification and description of aperture, which is commonly used in engineering practice, are given in Table 6.4.

6.2.5 Joint orientation

To define the orientation of discontinuities in rock mass, terms such as dip, dip direction and strike are used. *Dip* or dip angle represents the degree of inclination (Figure 6.7); it varies between 0° and 90°. *Dip direction* shows the orientation of the normal to the discontinuity plane against North (N). It is expressed by an angle from 0° to 360°. *Strike* is the alignment or run, which is the bearing of an imaginary horizontal line in the inclined plane (Figure 6.7).

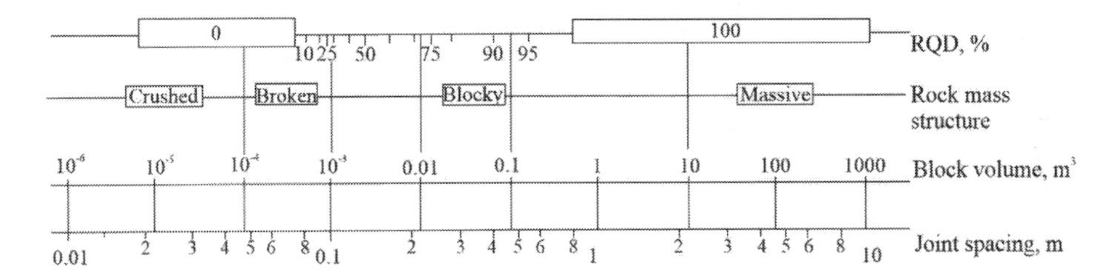

Figure 6.4 Correlations between the block volume, joint spacing and RQD (after Palmström, 1982)

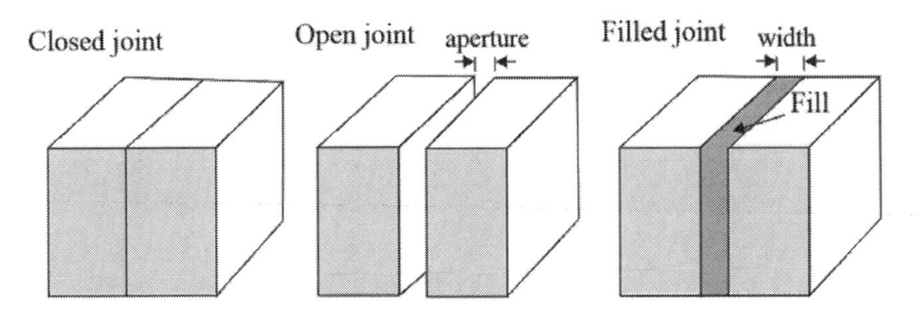

Figure 6.5 Types of joints: closed, open, and filled

Figure 6.6 Joints are filled with quartz

Table 6.4 Aperture in rock and its description (after Barton, 1978)

Aperture (mm)	*Description*	
< 0.1	Very tight	Closed features
0.1–0.25	Tight	
0.25–0.5	Partly open	
0.5–2.5	Open	Gapped features
2.5–10	Moderately wide	
> 10	Wide	
10–100	Very wide	Open features
100–1000	Extremely wide	
> 1000	Cavernous	

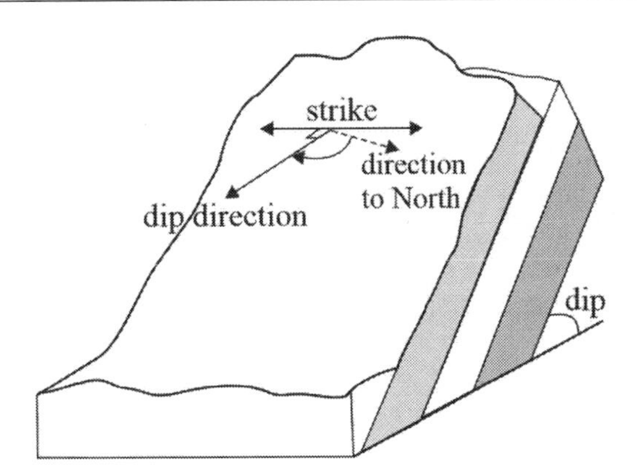

Figure 6.7 Joint orientations: dip, dip direction and strike

Dip direction and strike direction are always perpendicular to one another. The joint dip and dip direction are measured in the field by means of a geologic compass. In rock mechanics, dip direction/dip format is generally used to describe the orientation of discontinuities. For example, 210/35 or 030/35 (dip directions always have three digits).

The line of intersection between two planes is described using *plunge* and *trend*. The plunge is the inclination of the line to horizontal (similar to *dip*), while the trend (similar to *dip direction*) is the horizontal direction of the line measured clockwise from the north.

Orientation of a joint plane (dip and dip direction) can be analyzed graphically by means of an equatorial stereonet (Figure 6.8), in which joints are represented by either great circles or poles. The latitude lines define the dip direction while the longitude lines account for the dip. The center of the stereonet measures a dip of 90° while the perimeter line has a dip of 0°.

6.2.6 Project work: projections of major joint sets

Joint mapping of the rock outcrop at the project site (Point R in Figure 2.1) revealed three major joint sets whose dip and dip directions are summarized in Table 6.5. The following joint counts were made normal to each set: Joint set 1: 8 joints per 1 m; Joint set 2: 10 joints per 2 m; Joint set 3: 21 joints per 3 m. Estimate the volumetric joint count and RQD.

First, the data regarding the number of joints will be analyzed. The volumetric joint count (joints/m^3) is the sum of the number of joints per meter for each joint set (Equation 6.5).

$$J_v = \frac{1}{S_1} + \frac{1}{S_2} + \frac{1}{S_3} \tag{6.5}$$

where, S_1, S_2 and S_3 are measured in meters.

$$J_v = \frac{8}{1} + \frac{10}{2} + \frac{21}{3} = 20 \textit{ joints per } m^3$$

This rock mass can be classified as small-sized blocks (Table 6.3).

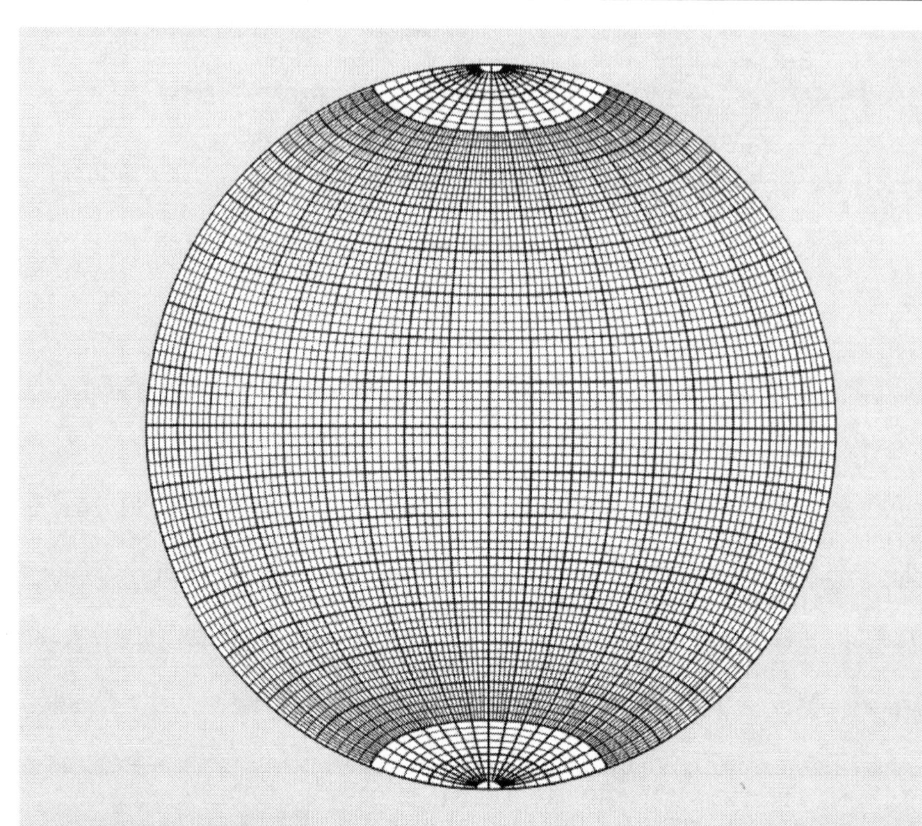

Figure 6.8 Equatorial stereonet

Table 6.5 Orientation (in degrees) of three major joint sets at Point R in Figure 2.1

	Dip	*Dip direction*
Joint set 1	24	125
Joint set 2	78	255
Joint set 3	35	040

The rock mass RQD can be estimated using Equation 6.4 as follows:

$$\mathrm{RQD}(\%) = 115 - 3.3 \cdot 20 = 49\%$$

The data on joint orientation will be analyzed and drawn by means of a stereonet. Tracing paper, an equatorial stereonet and a pin will be used. Follow these instructions to draw a great circle for Joint set 1 (24/125):

1. Place the tracing paper over the stereonet and fix it with the pin in the center.
2. Trace the circumference of the net and mark North on the tracing paper.
3. Along the perimeter, count the dip direction of 125° and mark it (Point A in Figure 6.9) on the tracing paper.

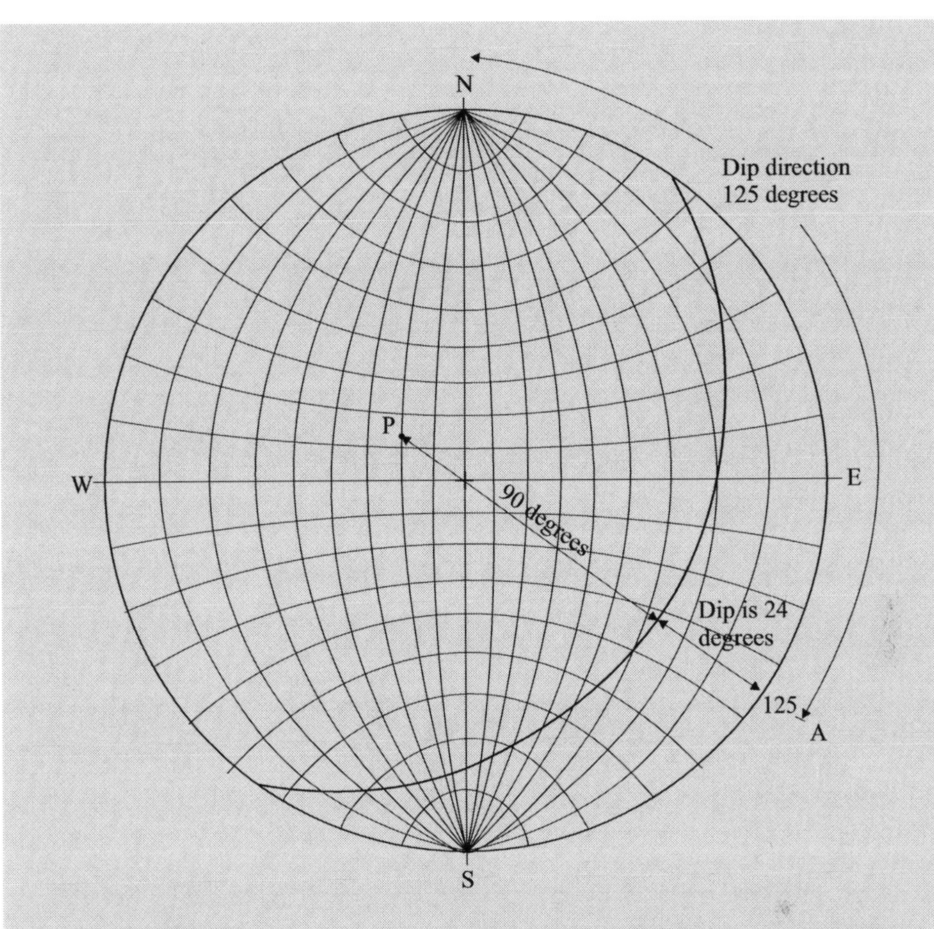

Figure 6.9 A great circle that represents a joint plane with the dip of 24° and the dip direction of 125°

4. Rotate the tracing paper (counterclockwise) so that the mark A is aligned with the E–W axis. This straight line (E–W axis) will be used to count the dip.
5. Starting from the perimeter, count 24° and trace the meridional circle corresponding to 24°. This is the great circle representing a joint plane of 24/125.
6. Mark the pole P on the tracing paper by counting 24° from the center.
7. Rotate the tracing paper back to its original position so that the North mark on the tracing paper coincides with the North on the equatorial stereonet underneath.

The great circles for three joint sets from Table 6.5 are given in Figure 6.10.

When two planes intersect, they form a line that can be represented in the stereonet as point. In the following procedure, the orientation (plunge and trend) for the point of intersection between Joint set 1 and Joint set 3 will be determined (Figure 6.11).

1. Draw a line from the center of the circle through the point of intersection and extend it so that it crosses the circumference (Point A in Figure 6.11).
2. Measure the trend, which equals 96°.

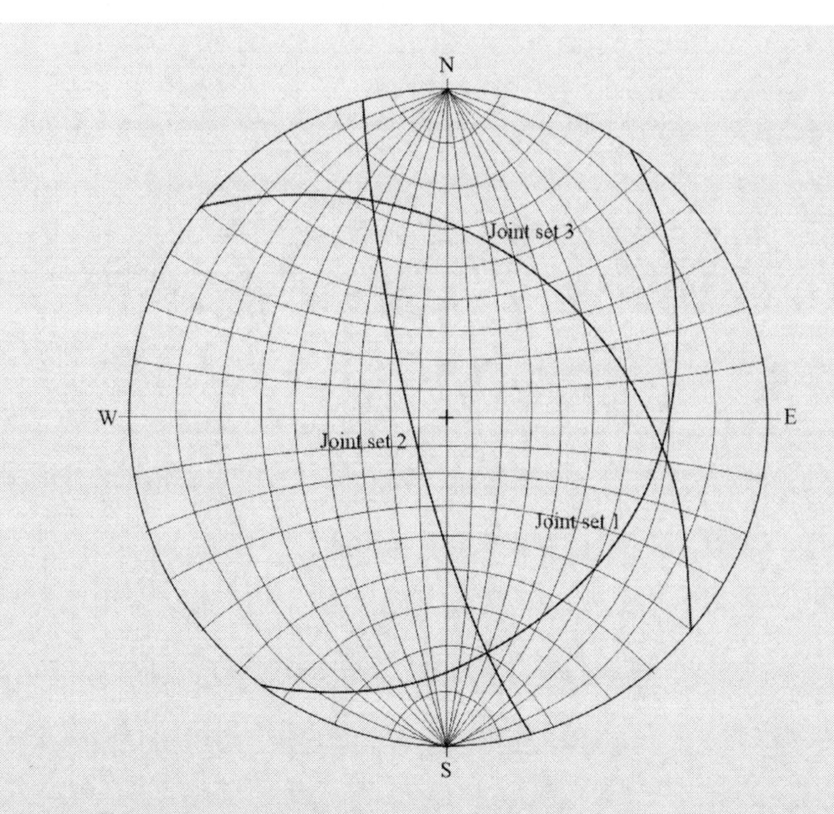

Figure 6.10 Great circles for three joint sets from Table 6.5

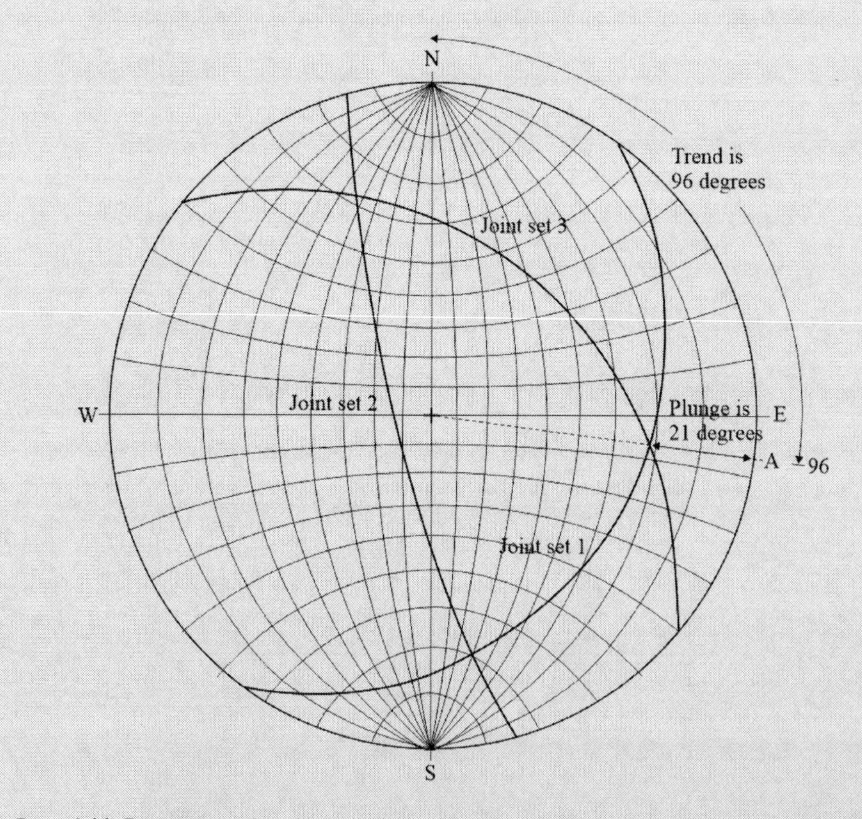

Figure 6.11 Determining the plunge and trend for the point of intersection between Joint set 1 and Joint set 3

Table 6.6 Trend and plunge of intersection points (in degrees)

Intersection	*Trend*	*Plunge*
Joint set 1/Joint set 2	169	17
Joint set 2/Joint set 3	340	20
Joint set 1/Joint set 3	096	21

3. Rotate the tracing paper until Point A lies on the E–W line of the equatorial stereonet.
4. Measure the plunge as 21°.

The plunge and trend for each point of intersection between the three major joint sets are given in Table 6.6.

Question: *When we deal with numerous discontinuities of large-sized rock mass, do we need to identify dip and dip direction for each joint?*
Answer: Yes, and it takes time to accomplish this task. However, there are new technologies available that can minimize the effort and time (Kim et al., 2015a,b). For example, photogrammetry, a remote sensing technique, is used to create a 3-D model of rock slopes (typically for slopes without soil layers and vegetation). It also allows to obtain the dip and dip direction of any joint from some distance, making field work safer (Gratchev et al., 2013; Kim et al., 2016).

6.3 Problems for practice

Problem 6.1: Estimate the RQD value of rock mass if the volumetric joint count is 7.6 joints/m^3.

Solution:

$$\text{RQD} = 115 - 3.3J_v = 115 - 3.3 \cdot 7.6 = 89.9(\%)$$

Problem 6.2: What is the joint frequency of jointed rock mass if a joint survey using a scanline approach (1 m long) produced an average joint spacing of about 5.8 cm?

Solution:

Joint frequency (λ) is related to the joint spacing (s_j) as expressed in Equation 6.2. Considering this,

$$\lambda = \frac{1}{s_j} = \frac{1}{0.058} \approx 17\ Joints/m$$

Problem 6.3: Using a scanline approach, the following measurements between the major joints were made on the face of a rock slope (Table 6.7). The first joint was at 0 cm while the last joint was at 100 cm. Estimate the volumetric joint count (J_v) for this part of the rock mass.

Table 6.7 The distance between the adjacent joints (in centimeters)

0	3.3	18.8	20.7	33.1	35.3	57.1	58.2	65.1	86.3	87.2	93.5	97.6	100

Solution:

We will measure the distance between the adjoining joints, and sum up the ones that are greater than 10 cm. There are four sections along the scanline that meet this requirement:

1. $18.8 - 3.3 = 15.5$ cm
2. $33.1 - 20.7 = 12.4$ cm
3. $57.1 - 35.3 = 21.8$ cm
4. $86.3 - 65.1 = 21.2$ cm

The RQD value of this rock mass can be estimated using Equation 6.1 as follows:

$$RQD = \frac{15.5 + 12.4 + 21.8 + 21.2}{100} \cdot 100\% \approx 70.9\%$$

Using the relationship between RQD and J_v (Equation 6.4), we will obtain

$$RQD = 115 - 3.3J_v$$

$$J_v \approx 13.4 \; Joints \; m^{-3}$$

6.4 Review quiz

1. Which of the following geologic processes may cause the formation of joints in rocks?

   a) erosion
   b) changes in stresses
   c) tectonic movements
   d) all of the above (a–c)

2. Joint frequency (λ) is related to the joint spacing (s_j) as follows:

   a) $\lambda = s_j$
   b) $\lambda = 10s_j$
   c) $\lambda = 1/s_j$
   d) no relationship exists

3. To describe the orientation of joints, the values of 210/35 are used. This means:

   a) dip/dip direction
   b) strike/dip direction
   c) dip direction/strike
   d) dip direction/dip

4. For joints in rock mass, a dip direction and strike direction are always:

   a) perpendicular
   b) parallel
   c) there is no relationship between them

5. Photogrammetry is mostly used for

   a) rock slopes
   b) earth slopes
   c) slopes with vegetation
   d) all three (a–c)

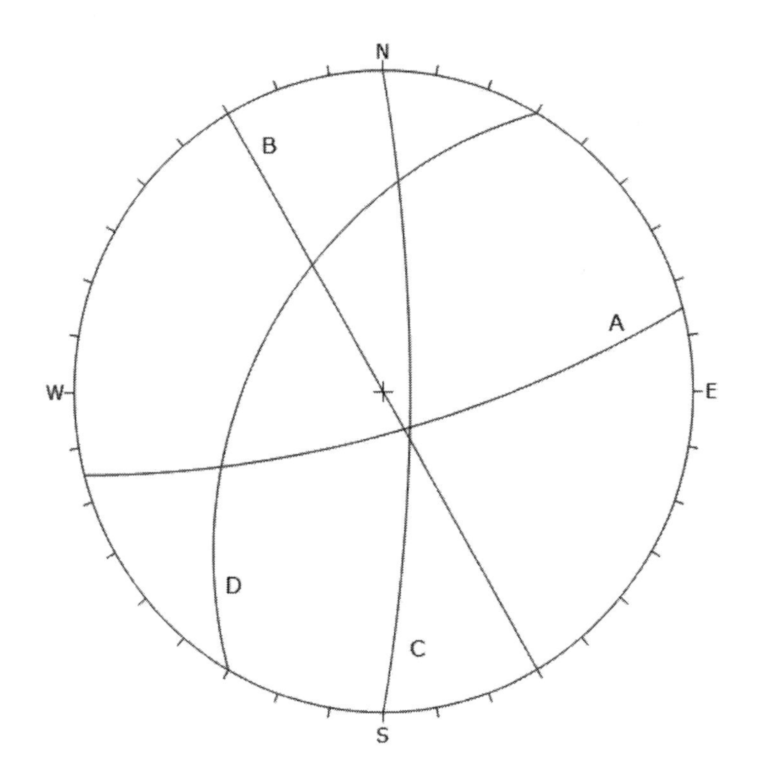

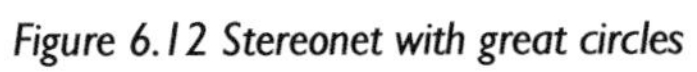
Figure 6.12 Stereonet with great circles

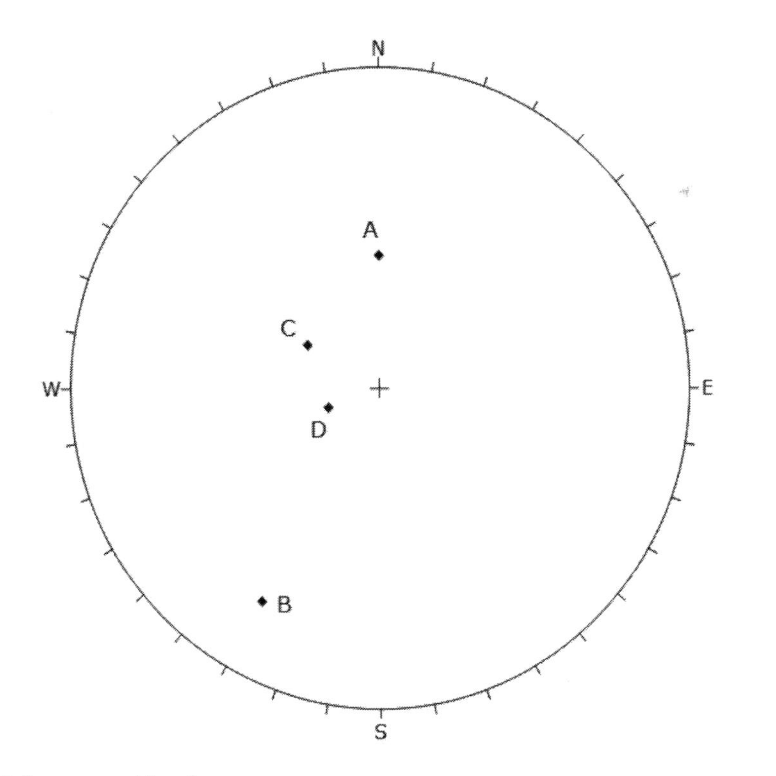

Figure 6.13 Stereonet with poles

6. Which of the four planes has the largest dip (Figure 6.12)?

   a) A b) B c) C d) D

7. Which of the four planes has the lowest dip (Figure 6.12)?

   a) A b) B c) C d) D

8. Which of the four planes has the lowest dip (Figure 6.13)?

   a) A b) B c) C d) D

Question: *Are there any tips on how to answer the stereonet questions above?*
Answer: Yes, you should remember that the farther the pole from the center, the larger the dip it has. However, it is opposite for great circles – the closer the projection of the great circle to the center, the greater dip it has.

9. Which of the following is the likely dip direction of plane A (Figure 6.13)?

   a) 0° b) 90° c) 180° d) 270°

Answers: 1. d 2. c 3. d 4. a 5. a 6. b 7. d 8. d 9. c

Chapter 7

Rock properties and laboratory data analysis

Project relevance: It is important to determine the engineering properties of rock, which can be used to assess the stability of rock mass. To obtain this information, a series of lab tests should be conducted, and the obtained data is interpreted and analyzed. This chapter introduces the basic properties of rock, discusses commonly used experimental procedures and explains how to interpret data from these tests. As part of the project work, the data from a series of triaxial, point load, unconfined compression and slake durability tests will be analyzed to determine the engineering properties of rock.

Question: *Rock specimens used in lab testing are of small size compared to the size of rock mass. Clearly, such specimens do not represent the jointed nature of rock mass. So, why do we still need to perform lab tests?*
Answer: Unfortunately, it is rather difficult to test large rock blocks due to technical issues and high cost of such experiments. For this reason, a different approach is used in which small-sized rock specimens are first tested in conventional apparatuses and the obtained results are then applied to jointed rock mass by using rock mass ratings. In this chapter, you will learn how to interpret results from lab tests, while in Chapter 9 you will learn how to apply these results to characterize the rock mass properties.

7.1 Rock properties

7.1.1 *Specific gravity*

Specific gravity (G) is defined as the ratio between the density of rock particles and the density of water (Equation 7.1). In other words, G is a number expressing how heavier the rock particles are compared to the same volume of water. The specific gravity is a dimensionless number, which is used to compute other rock properties such as porosity and density. Specific gravity of some common minerals is given in Table 7.1.

$$G = \frac{\rho_s}{\rho_w} \tag{7.1}$$

where, ρ_s is the density of solid particles, and ρ_w is the density of water.

Table 7.1 Specific gravity of common minerals

Mineral	*Specific gravity (G)*
Quartz	2.65
Feldspar	2.5–2.8
Calcite	2.7–3.0
Dolomite	2.8–3.0
Kaolinite	2.5–2.65
Montmorillonite	2–2.4
Muscovite (mica)	2.75–3.0

Question: *If rock consists of different minerals that have different values of specific gravity, what will be the absolute specific gravity of this rock?*
Answer: The absolute or true specific gravity of this rock will be the average specific gravity determined from a representative rock specimen.

7.1.2 Density and unit weight

Density (ρ) of rock is the ratio between the mass of rock sample (M) and its volume (V). If rock contains water, its total mass will include the mass of solid particles (M_s) and the mass of water (M_w) (Equation 7.2).

$$\rho = \frac{M}{V} = \frac{M_s + M_w}{V} \tag{7.2}$$

Question: *Igneous rocks seem to be denser than sedimentary rocks; what is the reason for this?*
Answer: Igneous as well as metamorphic rocks tend to contain heavier minerals compared to sedimentary rocks, which results in higher density values.

Dry density is the ratio of the mass of solids (M_s) to the total volume (V) of rock (Equation 7.3).

$$\rho_d = \frac{M_s}{V} \tag{7.3}$$

The unit weight (γ, in kN/m^3) is the ratio between the rock weight (W) and its volume (V). It is related to rock density (ρ), as shown in Equation 7.4.

$$\gamma = \frac{W}{V} = \rho \cdot g \tag{7.4}$$

where g is the gravitational acceleration ($\approx$ 9.81 m/s^2).

Question: *What is the typical range of unit weight for rocks?*
Answer: The unit weight depends on the type of rock minerals and rock structure; it may vary from 22 kN/m^3 to 30 kN/m^3. In some applications, when the unit weight is not known, it can be assumed to be 27 kN/m^3.

7.1.3 Project work: determination of rock density and unit weight

The density and unit weight of three rocks: sandstone, mudstone and andesite will be determined using the data from Table 2.1. The volume of cylindrical core samples is estimated as

$$V = \pi \cdot r^2 \cdot h$$

where r is the radius and h is the height of the rock specimen. The density is calculated using Equation 7.2, while the unit weight is computed using Equation 7.4. The obtained results are summarized in Table 7.2.

It is clear from Table 7.2 that the hard and fresh specimens of andesite have the greatest density (average of 2.65 g/cm³), while the relatively weak and weathered sandstone has the lowest average density of 2.07 g/cm³.

7.1.4 Porosity and void ratio

Porosity (n) expresses the amount of pore space in rock; it is defined as the ratio between the volume of voids (V_v) and the total volume (V) of rock (Equation 7.5).

$$n = \frac{V_v}{V} \tag{7.5}$$

Pores have a considerable effect on mechanical, especially strength properties of rock. A small amount of porosity can have a significant effect on rock deformation under stresses. Table 7.3 gives a range of porosity for some fresh rocks. It is evident from this data that fresh igneous rocks such as basalt and granite tend to have low values of porosity, while some soft sedimentary rocks (sandstone and shale) can have relatively high porosity values of 25%–30%.

Table 7.2 Density and unit weight of three rocks from the study area

Mass (g)	*Diameter (mm)*	*Height (mm)*	*Volume (cm³)*	*Density (g/cm³)*	*Unit weight (kN/m³)*
Sandstone					
435	51.3	101.2	209.1	2.08	20.4
440	51.9	101.6	214.8	2.05	20.1
431	51.2	101.1	208.0	2.07	20.3
Mudstone					
440	50.9	100.9	205.2	2.14	21.0
426	51.1	101.1	207.2	2.06	20.2
454	51.2	101.3	208.5	2.18	21.4
Andesite					
540	50.2	101.8	201.4	2.68	26.3
522	50.5	101.6	203.4	2.57	25.2
551	50.6	101.8	204.6	2.69	26.4

Table 7.3 Porosity values of some American rocks (after Griffith, 1937)

Rock	*Porosity (%)*
Basalt	0.2–22
Granite	1.0–2.9
Sandstone	1.6–26.4
Shale	20–50
Gneiss	0.3–1.6
Schist	10–30
Marble	0.6–0.8

Question: *When we visually examine core samples (e.g. fresh granite), it does not seem to have any pore space – is this correct?*
Answer: Some pore space may exist in fresh rocks (including basalt and granite) which is not visible to a naked eye. Porosity in crystalline rocks like granite is the result of internal stresses developed by modest changes in stress and temperature.

Void ratio (e) indicates the amount of voids in rock, and it is defined as the ratio of the volume of void space (V_v) to the volume of solids (V_s) (Equation 7.6). Void ratio is often used for soil material, while porosity is commonly used to describe rock properties.

$$e = \frac{V_v}{V_s} = \frac{n}{1-n} \tag{7.6}$$

7.1.5 Water content, degree of saturation and hydraulic conductivity

The water (or moisture) content (w) indicates the amount of water in rock. It is the ratio of the mass of water (M_w) to the mass of solids (M_s) (Equation 7.7). The moisture content of many fresh rocks is typically less than 1%; however, some weathered or soft rock (e.g. sandstone) can have much higher values.

$$w = \frac{M_w}{M_s} \tag{7.7}$$

Degree of saturation (S) indicates how much water is present in the voids. It is defined as the ratio of the volume of water (V_w) to the volume of void space (V_v) (Equation 7.8).

$$S = \frac{V_w}{V_v} = \frac{w \cdot G \cdot (1-n)}{n} \tag{7.8}$$

It is noted that water can exist in rock mass but it is generally localized in intergranular pores, natural cavities and discontinuities (joints and fractures) as schematically shown in Figure 7.1. As most fresh rock types have very low porosity, their rock permeability is generally very low. Ranges of hydraulic conductivity for common rocks are summarized in Table 7.4.

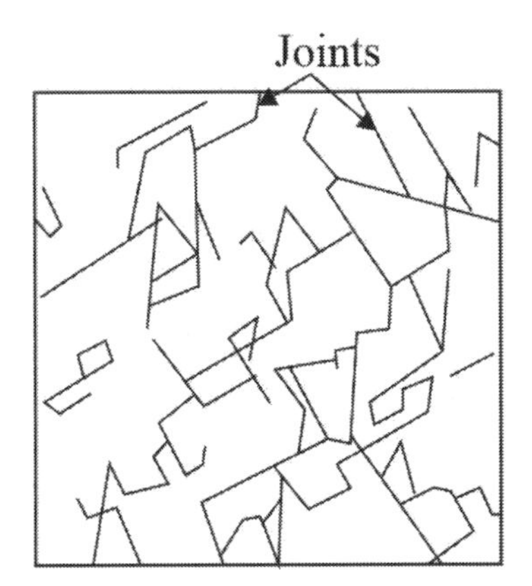

Figure 7.1 Joints in rock mass. Water can exist in joints; however, in fresh rocks joints may not be connected.

Table 7.4 Hydraulic conductivity of some rocks (from Atkinson, 2000)

Rock	*Hydraulic conductivity k (m/d)*
Sandstone	10^{-4}–10^{-1}
Shale	10^{-8}–10^{-4}
Unfractured igneous rocks	10^{-8}–10^{-5}
Fractured igneous rocks	10^{-3}–10
Karstic limestone	10^{-1}–10^{3}

Question: *Does this mean that rock should be treated as impermeable material?*
Answer: It depends on rock type and degree of weathering. Most fresh rocks can be impermeable; however, heavily weathered sedimentary rocks such as sandstone can be saturated with water similar to soil mass.

Unlike soil mass that has a ground water table, the distribution of water in rock mass depends on the size of pore space and degree to which they are interconnected.

7.2 Laboratory tests and data analysis

7.2.1 Unconfined compression test

The unconfined compression test is conducted on cylindrical rock specimens to obtain the strength properties of rock. Unlike soil material, rock generally fails at small strain (less than 1%), indicating brittle behavior. However, some soft rock (including shale and sandstone) can have relatively high strains at failure (ductile behavior).

Question: *Are there any criteria that define brittle and ductile rocks?*
Answer: According to Handin (1966), brittle rocks have axial strain of less than 1% at peak load, while ductile rocks display axial strain of more than 10% (Table 7.5).

Table 7.5 Difference in axial strain for brittle and ductile rocks (after Handin, 1966)

Rock behavior	*Axial strain (%)*
Very brittle	< 1
Brittle	1–5
Moderately brittle	2–8
Moderately ductile	5–10
Ductile	> 10

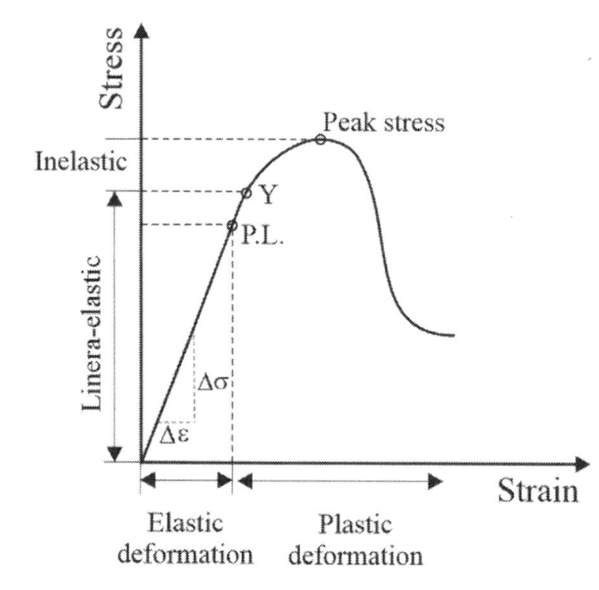

Figure 7.2 Typical stress (σ)-strain (ε) diagram for rocks. Y – yield point, P.L. – proportional limit.

7.2.2 Stress-strain diagram for rocks

Typical stress-strain behavior of rock under loads is shown in Figure 7.2. Engineering practice indicates that rock tends to fail or fracture at the proportional limit (point P.L. in Figure 7.2) of elasticity or somewhat beyond it very near the yield stress (point Y in Figure 7.2). As the applied load exceeds the yield point, permanent (plastic) deformations appear before the specimen reaches its peak strength.

Rocks that fail with no previous deformations are referred to as brittle. Rocks that develop plastic deformations before failure are considered to be ductile and fail by ductile rupture.

As most rocks are brittle, the plasticity domain of such rocks is relatively small. For such rocks, the stress is proportional to strain and the unconfined compression test typically ends in abrupt failure. Such materials are referred to as linear-elastic. The terms "quasi-elastic" and "semi-elastic" are used for rocks that exhibit a nearly linear stress-strain relationship to the point of failure.

Question: *Are rocks elastic or plastic material?*
Answer: Elasticity is a universal property of an ideal material, which deforms under stresses and recovers completely after the stress removal. Depending on how closely rocks approximate the ideal material, the concept of elasticity applies also to rocks. However, some soft rocks (e.g., sandstone) can also exhibit some plastic (permanent) deformations.

7.2.3 Modulus of elasticity and Poisson's ratio

According to Hooke's law, the stress σ (or $\Delta\sigma$) is proportional to strain ε (or $\Delta\varepsilon$), as shown in Equation 7.9.

$$\Delta\sigma = E \cdot \Delta\varepsilon \tag{7.9}$$

where E is the Young's modulus of elasticity. For most rock material, the proportional limit (point P.L. in Figure 7.2) coincides with the yield point (point Y).

For engineering purposes, it is essential to know the modulus of elasticity so that it is possible to accurately estimate the rock deformation under various load conditions. The modulus of elasticity, which can be obtained experimentally from unconfined compression tests, varies with the rock type, porosity, rock texture and water content. Typical values of the modulus elasticity for common rocks are given in Table 7.6.

Question: *For construction purposes, is it possible to increase the modulus of elasticity?*
Answer: Yes, it can be increased considerably by grouting.

By measuring diametrical strains during loading, Poisson's ratio (ν) can be obtained. Laboratory testing indicates that Poisson's ratio for most rocks varies from 0.15 to 0.35. Table 7.6 gives typical values of Poisson's ratio (ν) obtained for different types of rock.

Question: *Some metamorphic rocks such as slate and schist are strongly anisotropic. What direction shall be used to apply stresses in strength tests?*
Answer: Some metamorphic rocks which are composed of parallel arrangements of flat minerals like mica and chlorite exhibit strength *anisotropy*; that is, variation of compressive strength in relation to the direction of the principal stresses. Strength anisotropy can be evaluated by laboratory testing of specimens drilled in different directions from oriented block samples.

Table 7.6 Elasticity modulus and Poisson's ratio of some rocks

Rock	*Young's modulus of elasticity* (× 10^{10} N/m^2)	*Poisson's ratio*
Basalt	5–11	0.22–0.25
Granite	2–7	0.1–0.2
Limestone	1–8	0.125
Sandstone	0.5–8	0.12
Shale	1–3	0.11
Gneiss	2–6	0.25
Marble	6–9	0.25–0.3

7.2.4 Creep

As rocks are not perfect material, they combine the characteristics of elasticity, plasticity and flow in different proportions. Plastic flow, in which rock deforms slowly in a continuous way, is of particular interest in rock mechanics. Creep is the time-dependent movement of rock under a sustained stress, as shown in Figure 7.3.

In rock creep, three major phases can be recognized (Figure 7.3):

1. Primary creep, in which strain increases at a decreasing rate;
2. Secondary or steady-state creep, in which the rate of strain is constant;
3. Tertiary or accelerated phase – the phase of failure. During this phase, the strain increases at an increasing rate until the specimen fails or ruptures.

Creep is common phenomena in foliated rocks, in which stress-related permanent deformations occur along the foliated planes, resulting in rock mass movement (for example, landslides).

7.2.5 Tensile strength test

The tensile strength of material is defined as the maximum tensile stress which material is capable of developing. Rocks are relatively weak in tension because of the discontinuities they may have. For quick estimates, it is commonly assumed that the tensile strength (σ_t) of rock is about 10% of its compressive strength (σ_c).

$$\sigma_t = 0.1\sigma_c \tag{7.10}$$

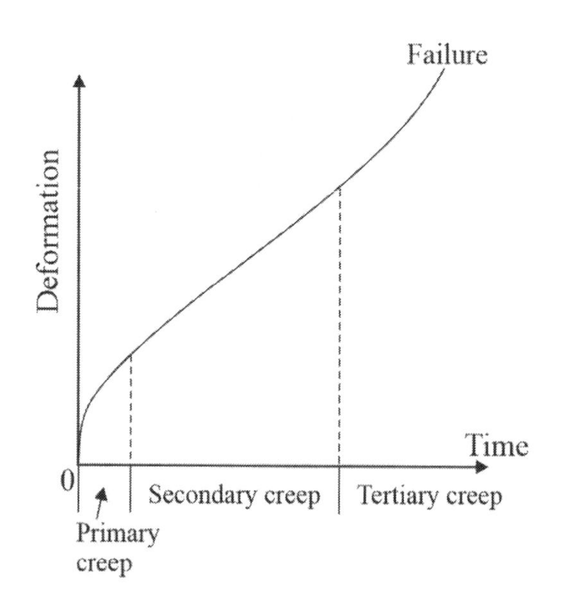

Figure 7.3 Time-dependent deformation at constant stress

Question: *Do we often use the tensile strength of rock in practice?*
Answer: Yes, knowledge of tensile strength is important in analyzing rock strength and stability of underground structures (e.g., the stability of roofs of underground openings in the tensile zone of the rock).

The Brazilian test is used to obtain the tensile strength of rock. In this test, a cylindrical rock specimen, lying on its side, is loaded diametrically with a compression load P (Figure 7.4). The tensile strength of the rock specimen in this test is estimated as:

$$\sigma_t = \frac{2P}{\pi \cdot d \cdot h} \tag{7.11}$$

where d is the specimen diameter and h is the height of the specimen.

Some typical values of rock strength (UCS and tensile strength) are summarized in Table 7.7.

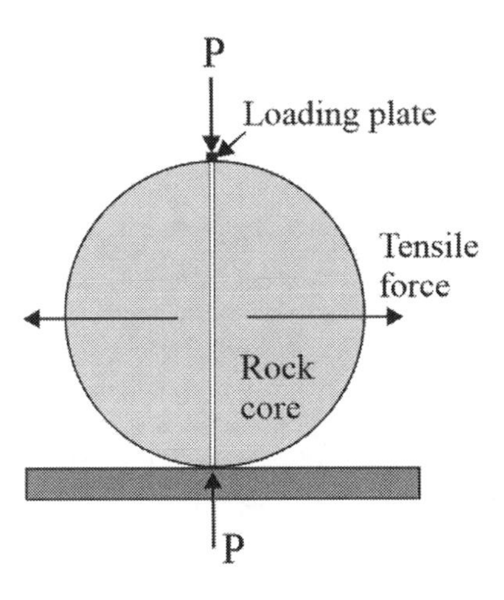

Figure 7.4 Schematic illustration of the Brazilian test to obtain the tensile strength of rock

Table 7.7 Strength characteristics of various rock types (after Jumikis, 1983)

Rock	*Unconfined compressive strength (MN/m^2)*	*Tensile strength (MN/m^2)*
Basalt	80–400	6–12
Granite	120–275	4–8
Limestone	4–196	1–7
Sandstone	20–167	4–25
Shale	21–160	30–110
Gneiss	78–245	4–7
Marble	49–177	5–8
Slate	98–196	7–20

7.2.6 Project work: estimation of unconfined compressive and tensile strength of andesite

Lab data from an unconfined compression test (Table 2.4), which was performed on a cylindrical specimen of andesite with a diameter of 50 mm, will be analyzed. The specimen area is

$$A = \pi \cdot r^2 = 3.14 \cdot 25^2 = 1.9625\, mm^2 \; or \; 0.0019625 m^2$$

Stress during the test is estimated as

$$\sigma = \frac{Load}{Area}$$

and it is given in Table 7.8 for each load increment.

The obtained data is plotted in Figure 7.5 as the stress against strain, and the peak stress (UCS) at failure is determined as 60 MPa.

Table 7.8 Data from unconfined compression test on andesite

Strain (%)	*Load (kN)*	*Stress (MPa)*
0	0	0
0.013	26	13.2
0.025	46	23.4
0.038	68	34.6
0.050	85	43.3
0.063	100	51.0
0.075	111	56.6
0.088	116	59.1
0.100	118	60.1
0.105	110	56.1

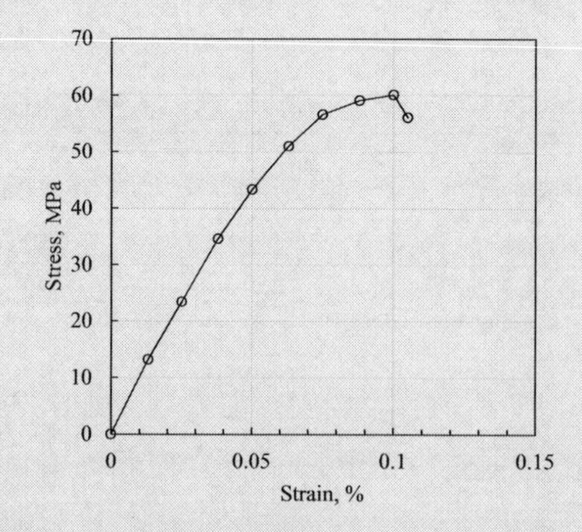

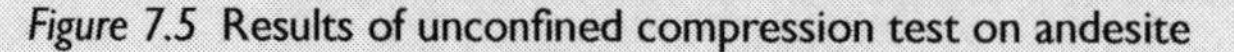

Figure 7.5 Results of unconfined compression test on andesite

Using Equation 7.10, the tensile strength can be estimated as follows:

$$\sigma_t = 0.1\sigma_c = 0.1 \cdot 60 = 6\ MPa$$

To obtain the modulus of elasticity, Equation 7.9 is used.

$$E = \frac{\Delta\sigma}{\Delta\varepsilon} = \frac{43.3}{0.0005} \approx 86600\ MPa\ or\ 86.6\ GPa$$

Question: *The size of specimens used in these tests seem to be very small. Will the specimen size affect the rock strength?*
Answer: Studies (Bieniawski, 1968) show that small-sized specimens tend to yield higher values of UCS than much larger-sized specimens of the same rock. However, larger specimens are more expensive to test, and they are mostly used for very important engineering projects.

7.2.7 Triaxial test

The triaxial compression test enables more accurate simulation of field stress conditions in the laboratory by considering the confining pressure ($\sigma_3 > 0$). However, it is a more expensive and time-consuming experiment compared to the unconfined compression. In triaxial tests, the specimen is confined under an equal all-around pressure of σ_3 (Figure 7.6). The axial stress (σ_1) is applied until the specimen fails. The deviator stress (q or $\Delta\sigma_d$), which equals the difference between the principal stresses (σ_1–σ_3), is used to describe the stress conditions at failure.

Data from a series of triaxial tests on the sandstone will be analyzed in Chapter 8 using different strength criteria.

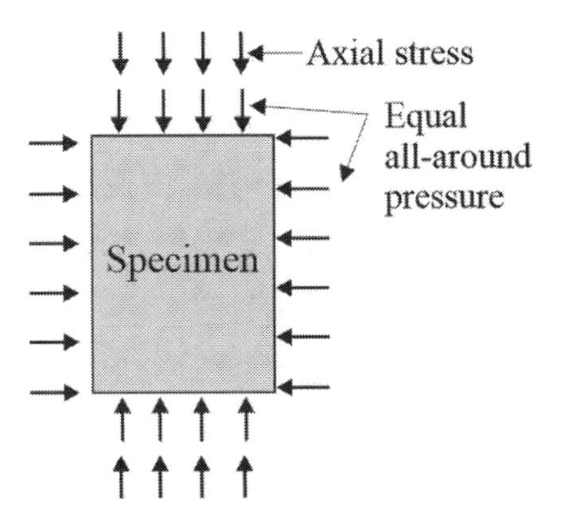

Figure 7.6 Stresses in a triaxial compression test

7.2.8 Project work: interpretation of point load test data

The relatively inexpensive and easy-to-perform point load test has become popular among engineers as it can be performed under field and laboratory conditions (Broch and

Franklin, 1972). It was developed as an index test for strength classification. In this test, a piece of rock is held between two conical platens and pressure *(P)* is increased until sample failure (Figure 7.7). The advantage of the point load test is that it can be carried out on irregular rock pieces in the field, following the standard test procedure (ASTM, 2008).

The size-corrected point load strength index $I_{s(50)}$ is correlated to UCS, as shown in Equation 7.12 (Broch and Franklin, 1972).

$$UCS \approx 24 \cdot I_{s(50)} \tag{7.12}$$

The data from a series of point load tests conducted on mudstone specimens (Table 2.3) will be analyzed. As the direction of load was *axial*, the equivalent core diameter, as shown in Figure 7.8, will be obtained using Equation 7.13.

$$D_e = \sqrt{\frac{4 \cdot A}{\pi}} \tag{7.13}$$

Using the recorded value of load at failure, the uncorrected point load strength is obtained as follows:

$$I_s = \frac{P \cdot 1000}{D_e^2}$$

It is normalized to the diameter of 50 mm using Equation 7.14.

$$I_{s(50)} = I_s \cdot \left(\frac{D}{50}\right)^{0.45} \tag{7.14}$$

Figure 7.7 Point load test setup

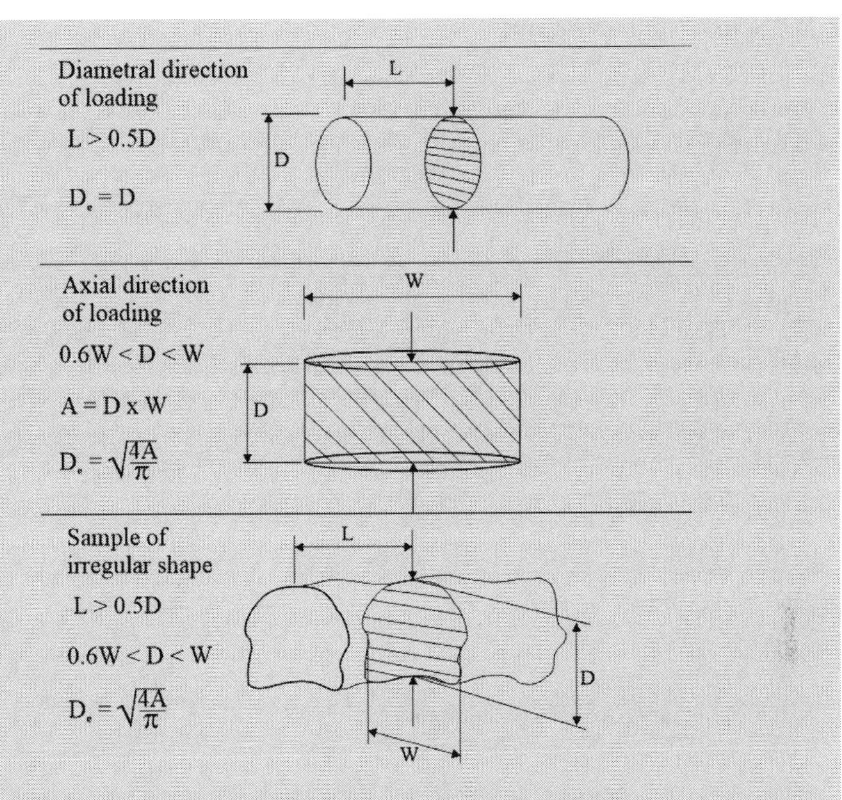

Figure 7.8 Shape proportions and equivalent core diameter of test specimens

Table 7.9 Point load data on mudstone and test interpretation

No.	Rock size		*Cross-sectional area*	*Equivalent core diameter*	*Load at failure*	*Uncorrected point load strength*	*Point load strength index*
	W	D	A = DW	D_e	P	I_s	$I_{s(50)}$
	mm	*mm*	*mm²*	*mm*	*kN*	*MPa*	*MPa*
1	50.9	37.2	1893.5	49.1	5.0	2.1	2.1
2	51.2	35.3	1807.4	48.0	4.1	1.8	1.7
3	50.6	33.6	1702.2	46.6	3.9	1.8	1.8

The obtained results for each test are given in Table 7.9. The average value of $I_{s(50)}$ is 1.87 while the UCS of mudstone can be estimated (Equation 7.12) as

$$UCS \approx 24 \cdot 1.87 \approx 44.8\ MPa$$

Question: *Is the multiplier (24) from Equation 7.12 used for all rocks?*

Answer: Studies have indicated that it cannot be applicable for all types of rocks, and thus, caution must be exercised for each particular rock. For example, the following multipliers are recommended for the Brisbane area of Australia (Look and Griffiths, 2001): tuff – 18, greywacke/argillite – 8, phyllite – 5.

7.2.9 Schmidt hammer test

The Schmidt hammer (Schmidt, 1951) (Figure 7.9) is a simple and portable device for testing rock hardness. It is a popular index test on rocks because its result (the rebound hardness R) is correlated with UCS. The hammer consists of a spring-loaded metal piston that is released when the plunger is pressed against the rock surface. The impact of the piston on the plunger transfers to the rock and the recovered energy from the rock is measured by the rebound height of the piston as the rebound value R.

This test is generally non-destructive and it can be done on any rock surface in the field or laboratory (ASTM, 2001). However, this test gives a wide range of values, and in order to obtain reliable results, at least ten measurements on the surface of interest are required. Standard practice recommends discarding the readings that differ from the average by more than 7 and averaging the rest. The average value of R is then used to estimate UCS (Figure 7.10).

The relationship between the rebound number (R), dry unit weight of rock (γ, in kN/m^3) and unconfined compressive strength (σ_c, in MPa) is given as follows:

$$\log_{10}(\sigma_c) = 0.00088 \cdot \gamma \cdot R + 1.01 \tag{7.15}$$

7.2.10 Slake durability test

Rock material is often used as fill in construction, and it is necessary to ensure that such rock can withstand long-term exposure to weathering, including the effect of water (Figure 7.11) (i.e., it is important to know rock durability).

Figure 7.9 The use of Schmidt hammer in practice

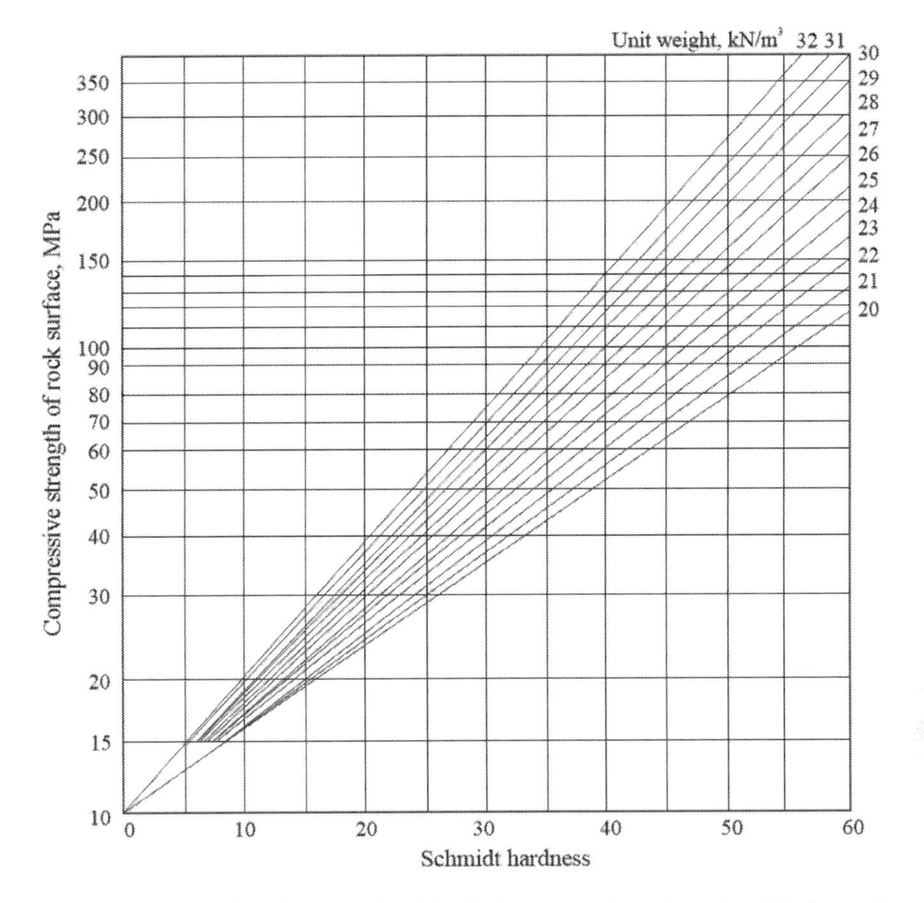

Figure 7.10 Correlation chart between the Schmidt hammer rebound number (R), dry rock unit weight and compressive strength (after Barton and Choubey, 1977)

Figure 7.11 Example of rock fill in construction where the daily wetting-drying process can undermine the strength of rock material over time

In practice, the slake durability index is used to indicate the resistance of rock to wetting and drying cycles. The slake durability test was proposed by Franklin and Chandra (1972); it is generally applied to weak rocks such as shales, mudstones and siltstones. The slake durability apparatus consists of a rotating sieve mesh drum which is immersed in a water bath. About ten rock lumps (each about 40–60 g) are placed in the drum and rotated for 10 minutes. During this time, disintegrated fragments can leave the drum through the 2 mm sieve mesh. The remaining fragments in the drum are dried and weighed. The dry mass of the rock fragments remaining in the drum after the second cycle is used to calculate the second-cycle slake durability index I_{d2} (in percent) as recommended by the relevant standard. For rock samples which are highly susceptible to slaking, I_{d2} is close to 0, while for very durable rock, I_{d2} is close to 100% (Table 7.10).

Question: *High durability means that rock can have a longer life; what is the life of common building stones?*

Answer: Some estimates include 20–40 years for limestone, 75–200 years for granite and a few centuries for gneiss. However, it strongly depends on factors related to the rock environment such as climate and atmosphere, resistance of rocks to frost action and rate of weathering.

Table 7.10 Durability classification based on slake durability index (after Franklin and Chandra, 1972)

Durability	I_{d1}	I_{d2}
Very high	> 99	98–100
High	98–99	95–98
Medium high	95–98	85–95
Medium	85–95	60–85
Low	60–85	30–60
Very low	< 60	0–30

7.2.11 Project work: determination of slake durability index for andesite and sandstone

We will estimate the slake durability index of sandstone and andesite and determine what rock is preferable as fill. The lab data for each rock is given in Table 7.11.

Table 7.11 Slake durability test data

Rock type	*Andesite*			*Sandstone*		
Test No.	*1*	2	3	*1*	2	3
Mass of tray (m_4), g	31.84	31.61	31.68	44.6	44.5	44.5
Mass of tray and dry sample (m_1), g	389.56	201.78	330.22	405.3	262.2	409.3
Mass of tray and dry sample after first cycle (m_2), g	388.04	197.94	326.08	350.3	230.4	363.3
Mass of tray and dry sample after second cycle (m_3), g	385.17	197.13	318.13	344.4	226.1	357.6
First cycle slake durability index, I_{d1}, %	99.6	97.7	98.6	84.75	85.39	87.39
Second cycle slake durability index, I_{d2}, %	98.8	97.3	96.0	83.12	83.42	85.83

Note that

$$I_{d1} = \frac{m_2 - m_4}{m_1 - m_4} \cdot 100\%$$

$$I_{d2} = \frac{m_3 - m_4}{m_1 - m_4} \cdot 100\%$$

The average value of I_{d2} ($\approx$ 97.4%) for the andesite is much higher than the one obtained for the sandstone ($\approx$ 84.1%), suggesting that the andesite is a better choice as rock fill.

7.3 Problems for practice

Problem 7.1: Determine the unconfined compressive strength (UCS) of rock based on the results of Schmidt hammer tests. The readings are 40, 37, 38, 43, 36, 44, 44, 40, 41 and 42. Rock density is 2.35 g/cm^3.

Solution:

The average rebound value equals 40.5. The unit weight of this rock equals

$$\gamma = 2.35 \cdot 9.81 \approx 23.1\ kN / m^3$$

From the chart in Figure 7.10, the unconfined compressive strength will be about 66 MPa.

Problem 7.2: The rock has density of 2.85 g/cm^3 and tensile strength of 8.9 MPa. Estimate the unconfined compressive strength (UCS) of this rock.

Solution:

Using Equation 7.10, the unconfined compressive strength can be found as

$$UCS = 10 \cdot \sigma_t = 10 \cdot 8.9 = 89\ MPa$$

Problem 7.3: Estimate the slake durability index I_{d2} for granite using the data given in Table 7.12.

Table 7.12 Results of slake durability tests

Measurements	*Mass (g)*
Mass of drum	971
Mass of drum + dry sample	1476
Mass of drum + dry sample after first cycle	1472
Mass of drum + dry sample after second cycle	1467
Mass of drum + dry sample after third cycle	1464

Solution:

The initial mass of dry specimen is

$$M = 1476 - 971 = 505\ g$$

The mass of dry specimen after the second cycle (M_{d2}) equals

$$M_{d2} = 1467 - 971 = 496\,g$$

The slake durability index for the second cycle (I_{d2}) will be

$$I_{d2} = \frac{496}{505} \cdot 100\% = 98.2\%$$

Problem 7.4: A series of point load tests on slightly weathered sandstone resulted in the average value of point load strength index ($I_{s(50)}$) of 1.12 MPa. Estimate the unconfined compressive strength (UCS) of this rock.

Solution:

From Equation 7.12, the unconfined compressive strength is obtained as

$$UCS = I_{s(50)} \cdot 24 = 1.12 \cdot 24 \approx 26.9\ MPa$$

Problem 7.5: A series of point load tests on slightly weathered rock resulted in the average value of point load strength index of 0.8 MPa. Estimate the tensile strength (in MPa) of this rock.

Solution:

$$UCS = I_{s(50)} \cdot 24 = 0.8 \cdot 24 \approx 19.2\ MPa$$

The tensile strength can be estimated using Equation 7.10:

$$\sigma_t \approx \frac{UCS}{10} = \frac{19.2}{10} = 1.92\ MPa$$

Problem 7.6: The average Schmidt hammer value is 25. Estimate the point load strength index (in MPa) of this rock.

Solution:

As there is no information about the rock unit weight, assume that it is 27 kN/m^3 (Section 7.1.2). From the Schmidt hammer chart (Figure 7.10), the unconfined compressive strength is UCS $\approx$ 40 MPa. Using Equation 7.10, the point load index strength is estimated as follows:

$$I_{s(50)} \approx \frac{UCS}{24} = \frac{40}{24} \approx 1.67\ MPa$$

7.4 Review quiz

1. What would be a typical range of porosity for fresh granite and gabbro?

   a) 0.1%–0.5% b) 4%–5% c) 8%–10% d) 12%–15%

2. The typical range for the uniaxial compressive strength of rock is

   a) 1–400 kPa b) 1–400 MPa c) 1–400 GPa d) none of these

3. What would be a typical range of unconfined compressive strength of fresh basalt?

   a) 1–5 MPa b) 10–15 MPa c) 20–25 MPa d) more than 100 MPa

4. Which of the following tests will measure the tensile strength of rocks?

   a) point load test b) unconfined compression test
   c) Brazilian test d) triaxial compression test

5. What is the empirical relationship between the unconfined compressive strength (UCS) and tensile strength (T)?

   a) UCS ≅ 10T b) UCS ≅ 24T c) UCS ≅ 35T d) UCS ≅ 5T

6. The behavior of extremely hard rocks in uniaxial compression tests after the peak strength is reached is typically described as ductile.

   a) True b) False

7. Which of the following rocks generally has greater hydraulic conductivity?

   a) shale b) sandstone
   c) fresh basalt d) karstic limestone

Answers: 1. a 2. b 3. d 4. c 5. a 6. b 7. d

Chapter 8

Stresses and failure criteria

Project relevance: During construction, when large loads are applied to rock, it may undergo deformations which undermine the overall stability of rock mass. To estimate the behavior of rock mass, it is important to know the level of stress acting on rock at a certain depth. This chapter will explain how to calculate the vertical stress and discuss the major failure criteria that engineers use to assess the rock strength. As part of the project work, the laboratory data obtained for the sandstone and mudstone will be analyzed to determine the strength characteristic of these rocks.

8.1 Stresses in rock mass

For rock mass (Figure 8.1), the vertical stress (σ_v) at a depth of z is given by the weight of the overlying material (Equation 8.1)

$$\sigma_v = \gamma \cdot z \tag{8.1}$$

where γ is the unit weight of the overlying rock material.

In rock mechanics, it is common to assume that the average unit weight of rock is 27 kN/m^3. Then, the average vertical stress at depth (z) can be estimated as

$$\sigma_v \approx 0.027 \cdot z \tag{8.2}$$

where σ_v is in MPa, and z is in meters.

Unlike soil mass, the horizontal stresses in rock mass are mostly due to tectonic activities, and they can be greater than the vertical stresses.

Question: *Similar to soil mechanics, can we still use the effective stress concept as shown in Equation 8.3?*

$$\sigma' = \sigma - u \tag{8.3}$$

where σ' is the effective stress, σ is the total stress, and u is the pore water pressure.

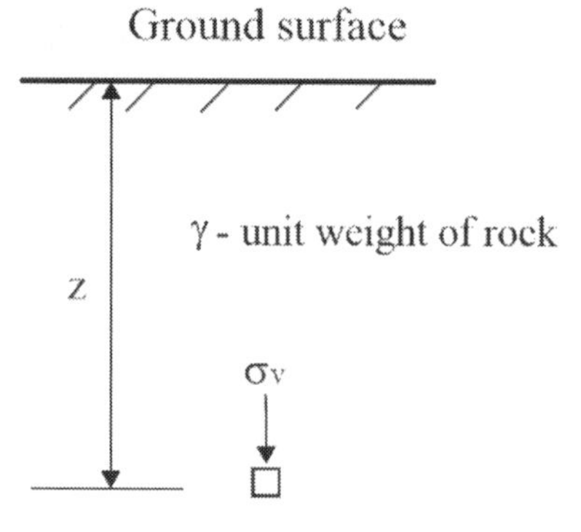

Figure 8.1 Determination of vertical stress in rock mass

Answer: Due to the nature of ground water in rock, the effective stress concept may not work for some rock mass where water is only present in non-connected fractures.

8.2 Failure criteria

This section will explain how to analyze lab data using different failure criteria. For triaxial tests, two major failure criteria, namely Mohr-Coulomb and Hoek-Brown, will be used to determine the strength characteristics of rock.

Question: *Are there any other criteria which we should be aware of?*
Answer: The Griffith strength criterion has successfully been used before to estimate the tensile strength of rock; however, it tends to underestimate the compressive strength. For this reason, it is not common these days.

8.2.1 Mohr-Coulomb strength criterion

The Mohr-Coulomb failure criterion is used for rock, in which failure occurs in shear. The shear strength is proportional to the normal stress acting on the shear plane and expressed as

$$\tau = \sigma \cdot \tan\phi + c \tag{8.4}$$

where, τ is the shear strength, c is the cohesion, and ϕ is the friction angle.

In rock mechanics, the knowledge of the principal stresses (σ_1 and σ_3) at failure as well as unconfined compressive strength (σ_c) of rock is extremely important. For this reason, the Mohr-Coulomb strength criterion (Equation 8.4) is modified to include these parameters (Equation 8.5).

$$\sigma_1 = 2c \cdot \tan\left(45 + \frac{\phi}{2}\right) + \sigma_3 \tan^2\left(45 + \frac{\phi}{2}\right) = \sigma_c + \sigma_3 \tan^2\left(45 + \frac{\phi}{2}\right) \tag{8.5}$$

8.2.2 Hoek-Brown failure criterion for intact rock

Hoek and Brown (1980) proposed that the effective principal stresses at failure can be related as

$$\sigma'_{1f} = \sigma'_{3f} + \sigma_{ci}\left(m_b \frac{\sigma'_{3f}}{\sigma_{ci}} + s\right)^{\alpha} \tag{8.6}$$

where m_b, s and α are variables which are equal to m_i, 1 and 0.5, respectively, for intact rock only. Considering this, Equation 8.6 can be rewritten as

$$\left(\sigma'_{1f} - \sigma'_{3f}\right)^2 = m_i \sigma_{ci} \sigma'_{3f} + \sigma_{ci}^2 \tag{8.7}$$

Values of σ_{ci} (unconfined compressive strength of intact rock) and m_i can be determined from a series of triaxial tests when the obtained test results are plotted as $(\sigma'_{1f} - \sigma'_{3f})^2$ against σ'_{3f}. The tensile strength (σ_{ti}) of rock can be estimated as expressed in Equation 8.8:

$$\left[\frac{\sigma_{ti}}{\sigma_{ci}}\right] = -\frac{\left(\sqrt{m_i^2 + 4s}\right) - m_i}{2} \tag{8.8}$$

8.3 Project work: lab data analysis and rock properties

The results of triaxial tests on the sandstone (Table 2.2) will be analyzed using the Mohr-Coulomb failure criterion. The data from these tests is plotted in Figure 8.2 in terms of the principal stresses at failure. Using Microsoft Excel, the linear trendline for the experimental data as described by Equation 8.9 is obtained.

$$y = 3.76x + 30 \tag{8.9}$$

By comparing Equation 8.9 with the Mohr-Coulomb failure criterion (Equation 8.10)

$$\sigma_{1f} = \sigma_c + \sigma_{3f} tan^2\left(45 + \frac{\phi}{2}\right) \tag{8.10}$$

It can be noted that $y = \sigma_{1f}$ and $x = \sigma_{3f}$, resulting in

$$\sigma_c = 30\ MPa$$

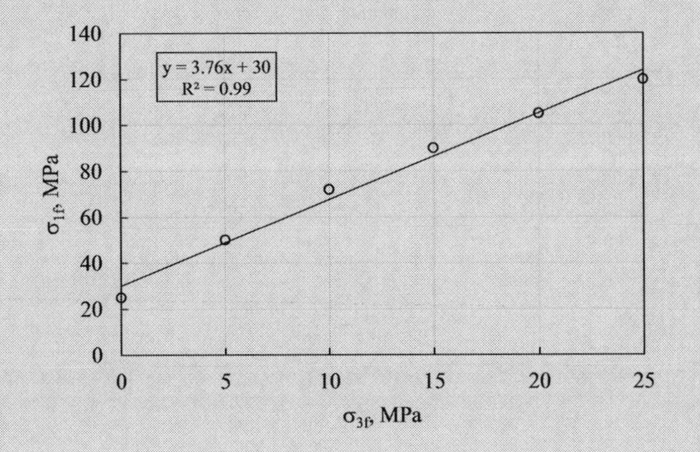

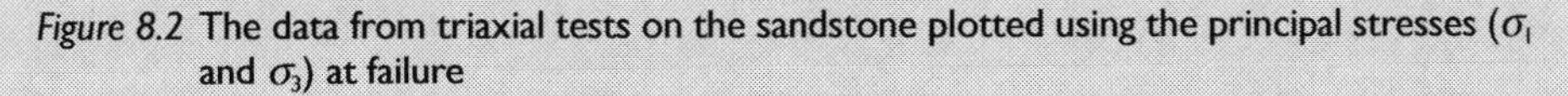
Figure 8.2 The data from triaxial tests on the sandstone plotted using the principal stresses (σ_1 and σ_3) at failure

The friction angle is estimated from the following equation:

$$tan^2\left(45+\frac{\phi}{2}\right)=3.76$$

Solving which, the friction angle is obtained as

$$\phi\approx 35.4^{\circ}$$

Substituting the friction angle in Equation 8.5, the rock cohesion can be found

$$\sigma_1 = 2c\cdot \tan\left(45+\frac{35.4}{2}\right)+\sigma_3 \tan^2\left(45+\frac{35.4}{2}\right)$$

The cohesion equals 7.7 MPa.

Now the same lab data will be analyzed using the Hoek-Brown failure criterion. The triaxial test results are replotted in Figure 8.3 in terms of the difference in the principal stresses, and the trendline equation (Equation 8.11) is obtained using Excel functions.

$$y = 339.3x + 486.77 \tag{8.11}$$

By comparison with the Hoek-Brown (Equation 8.12)

$$\left(\sigma'_{1f}-\sigma'_{3f}\right)^2 = m_i\,\sigma_{ci}\,\sigma'_{3f}+\sigma^2_{ci} \tag{8.12}$$

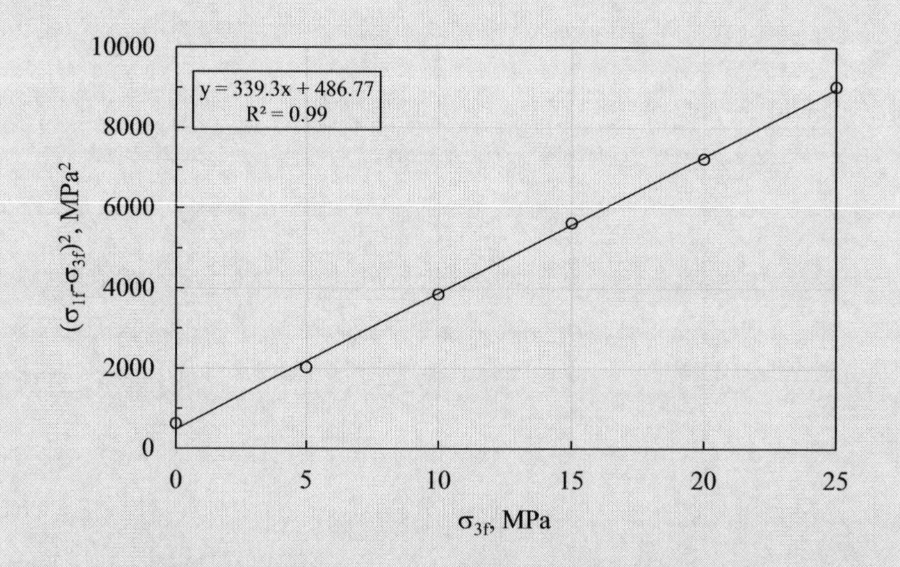

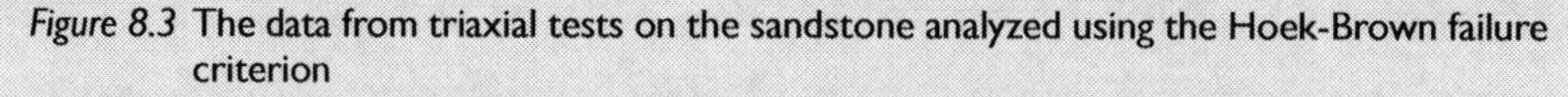

Figure 8.3 The data from triaxial tests on the sandstone analyzed using the Hoek-Brown failure criterion

the following results can be concluded:

$$y = \left(\sigma'_{1f} - \sigma'_{3f}\right)^2$$
$$x = \sigma'_{3f}$$
$$\sigma_{ci}^2 = 486.7$$
$$\sigma_{ci} \approx 22.1\ MPa$$

From

$$m_i \cdot \sigma_{ci} = 339.3$$

The m_i value can be obtained as

$$m_i = \frac{339.3}{22.1} \approx 15.4$$

To estimate the tensile strength of this rock, Equation 8.8 will be used as follows:

$$\left[\frac{\sigma_{ti}}{\sigma_{ci}}\right] = -\frac{\left(\sqrt{m_i^2 + 4s}\right) - m_i}{2}$$
$$\left[\frac{\sigma_{ti}}{22.1}\right] = -\frac{\left(\sqrt{15.4^2 + 4 \cdot 1}\right) - 15.4}{2}$$
$$\sigma_{ti} \approx -1.4\ MPa$$

Question: *The value of σ_{ci} obtained using the Hoek-Brown criterion does not match the one from the Mohr-Coulomb failure envelope; which one is correct?*

Answer: Engineering practice shows that both criteria provide good estimations of the unconfined compressive strength of rock; however, the Mohr-Coulomb envelope tends to overestimate the tensile strength of rock, as schematically shown in Figure 8.4.

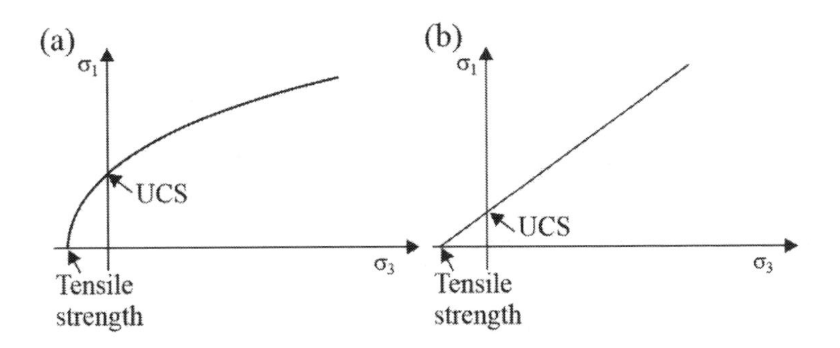

Figure 8.4 Comparisons between the Hoek-Brown failure criterion (a), and Mohr-Coulomb failure criterion (b). UCS – unconfined (uniaxial) compressive strength.

8.4 The Barton shear strength criterion for jointed rocks

The Barton criterion (Barton, 1973) is used to estimate the shear strength of jointed rock mass (not intact rocks). It considers the joint properties such as the joint roughness coefficient (JRC), joint wall strength (JCS), and the residual shear strength (ϕ_r) between two joints. It is commonly used to assess the stability of rock slopes (Kim et al., 2013; Cui et al., 2019) (Figure 8.5a) and failure of rock blocks along the defined joint (Figure 8.5b).

The shear strength (τ) of jointed rock mass is estimated as given in Equation 8.13.

$$\tau = \sigma_n \tan\left(\phi_r + JRC \log_{10}\left(\frac{JCS}{\sigma_n}\right)\right) \tag{8.13}$$

where σ_n is the normal stress acting on the joint.

If the joints are completely unweathered, then JCS equals the unconfined compression strength of the unweathered rock. The joint roughness coefficient is obtained by examining

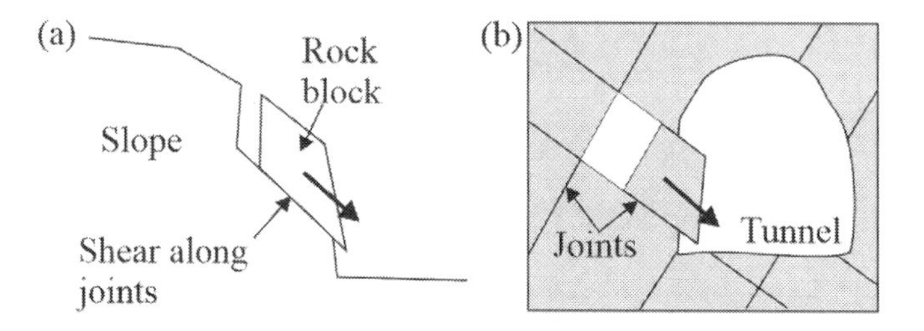

Figure 8.5 Examples of rock sliding along the joint: (a) slope stability issues and (b) block sliding in the tunnel wall

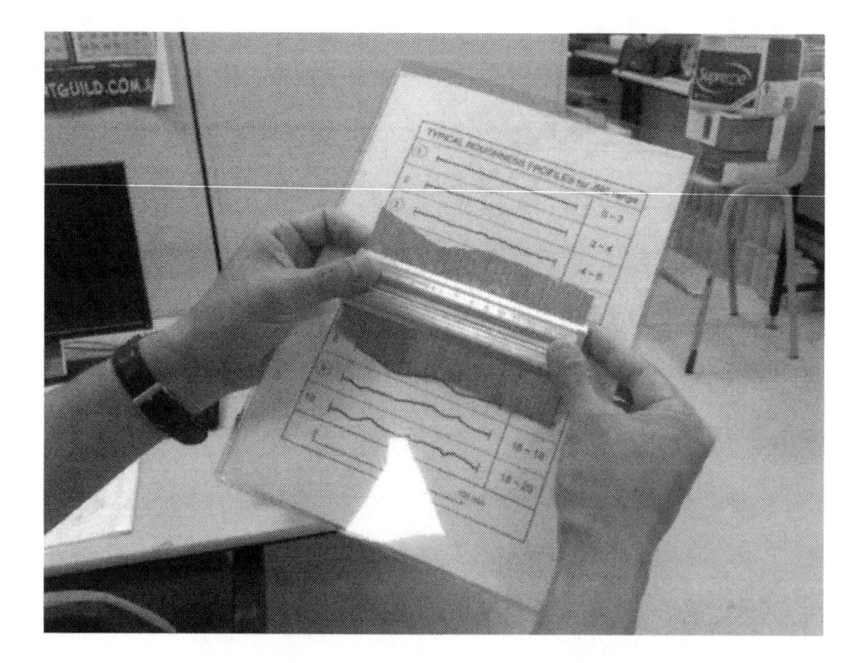

Figure 8.6 Estimating JRC of rock surface by means of a Barton's comb

the joint surface using a Barton's comb (Figure 8.6) and comparing it with the standard rock profiles proposed by Barton and Choubey (1977).

Examples of such joint surface profiles with the assigned JRC values are given in Figure 8.7. The joint roughness coefficient (JRC) is used to describe the surface irregularities; it ranges from 0 for smooth surfaces to 20 for rough stepped and undulating surfaces.

The friction between two joints can be estimated by means of a tilt test (Figure 8.8) using smooth unweathered rock surfaces. In this test, the top half of the jointed rock specimen

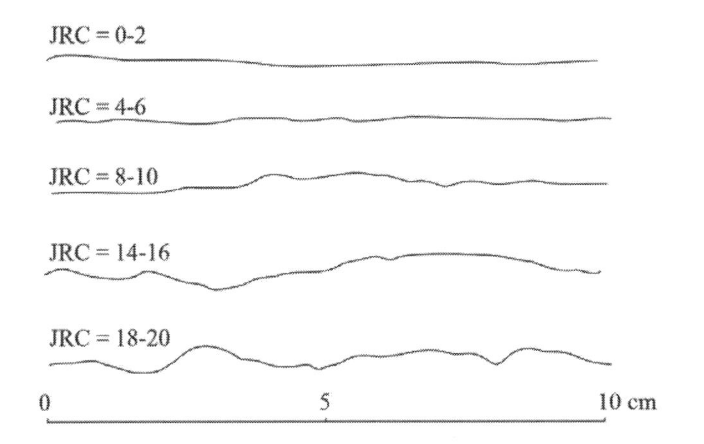

Figure 8.7 Examples of surface roughness profiles and corresponding joint roughness coefficients after Barton and Choubey (1977)

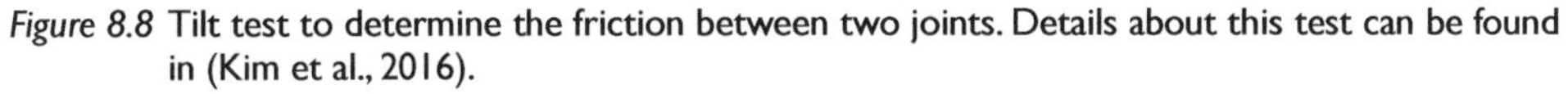

Figure 8.8 Tilt test to determine the friction between two joints. Details about this test can be found in (Kim et al., 2016).

slides as the tilting angle increases. Previous research (Barton and Choubey, 1977) indicates that although the basic friction angle depends on the rock origin, for many rock types, it varies from 26° to 35°.

8.5 Project work: Barton shear strength criterion for the jointed mudstone

The shear strength of jointed mudstone will be estimated using the Barton shear strength criterion. The rock core, which was collected from a depth of 7 m, had density of 2.7 g/cm^3, the JRC value varied from 16 to 18, and the residual friction angle was 31°. A series of Schmidt hammer tests on the rock joint surfaces resulted in an average value of Schmidt hammer rebound number of 35. Estimate the shear strength of jointed mudstone at a depth of 7 m.

The unit weight of this mudstone specimen is

$$\gamma = \rho \cdot g = 2.7 \cdot 9.81 \approx 26.5\ kN\ m^{-3}$$

From the chart in Figure 7.10, UCS (= JCS) will be about 68 MPa.

The normal stress at a depth of 7 m equals

$$\sigma = 26.5 \cdot 7 = 185.5\ k\ Pa = 0.186\ MPa$$

From Equation 8.13, the shear strength of the jointed mudstone will be

$$\tau = 0.186 \cdot \tan\left(31 + 17 \cdot \log\left(\frac{68}{0.186}\right)\right) \approx 0.674\ MPa$$

8.6 Problems for practice

Problem 8.1: Estimate an average vertical stress in rock mass at a depth of 500 m.

Solution:

As no data on rock density is provided, Equation 8.2 will be used to estimate the vertical stress at a depth of 500 m

$$\sigma = 0.027 \cdot 500 = 13.5\ MPa$$

Problem 8.2: A tunnel will be built in jointed granite at a depth of 90 m. The following information is provided: the rock density is 2.9 g/cm^3, the JRC of joints is in the range of 16–18, Schmidt hammer tests on the rock joint surfaces resulted in UCS = 85 MPa, and the residual friction angle was about 31°. Using the Barton criterion, estimate the shear strength of rocks at the tunnel's depth.

Solution:

The unit weight of granite is

$$\gamma = \rho \cdot g = 2.9 \cdot 9.81 \approx 28.45\ kN\ m^{-3}$$

The normal stress at the tunnel's depth (z = 90 m) equals

$$\sigma_n = \gamma \cdot z = 28.45 \cdot 90 \approx 2.56\ MPa$$

Using Equation 8.13, the shear strength of jointed granite will be

$$\tau = 2.56 \cdot \tan(31 + 17 \cdot \log_{10}\left(\frac{85}{2.56}\right)) \approx 3.92\ MPa$$

Problem 8.3: Triaxial tests were carried out on 50 mm diameter rock cores and the following data (Table 8.1) were obtained for the principal stresses at failure. This data is also plotted in Figures 8.9 and 8.10.

1. Estimate the tensile strength (in MPa) of this rock using the Hoek-Brown failure criteria.
2. Estimate the friction angle and cohesion using the Mohr-Coulomb failure criteria.

Table 8.1 Results of triaxial tests

σ_{3f} (MPa)	0	5	10	15	20	25
σ_{1f} (MPa)	25	60	77	105	125	140

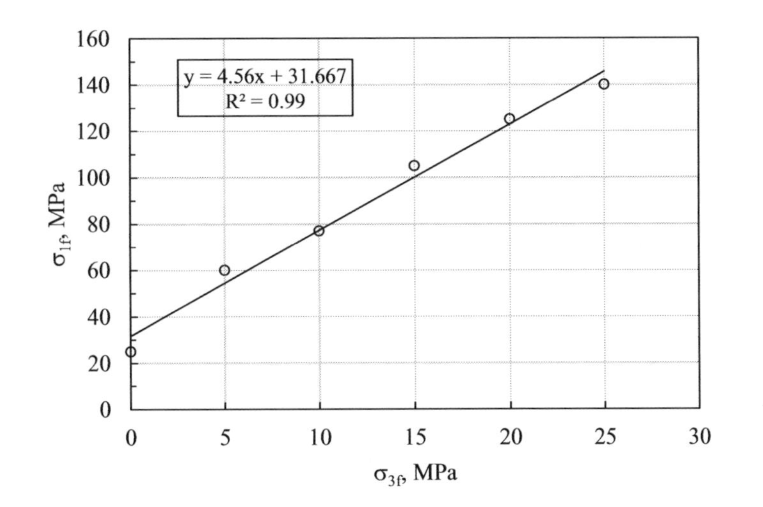

Figure 8.9 Results from triaxial tests

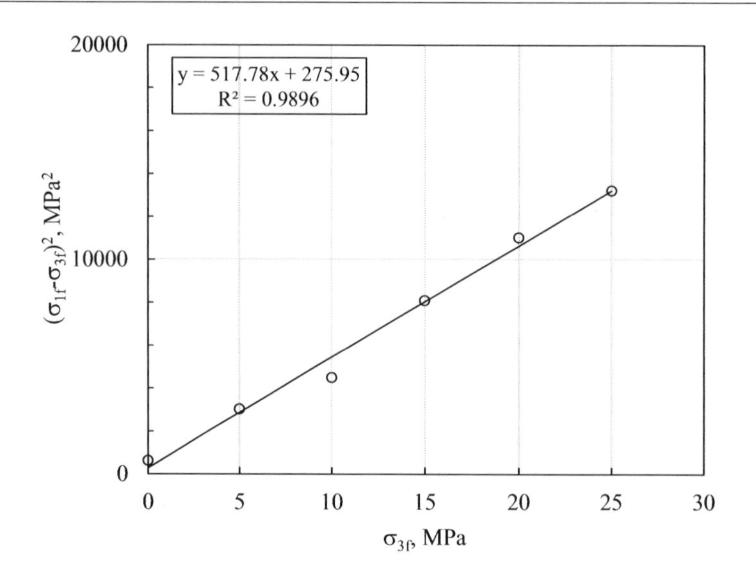

Figure 8.10 Results from triaxial tests

Solution:

1. Using the Hoek-Brown failure criteria (Figure 8.10), it can be concluded that

$$\sigma_{ci}^2 = 275.95$$

$$\sigma_{ci} \approx 16.6\ MPa$$

$$m_i = \frac{517.78}{16.6} \approx 31.2$$

Using Equation 8.8, the tensile strength is obtained as follows:

$$\left[\frac{\sigma_{ti}}{16.6}\right] = -\frac{\left(\sqrt{31.2^2 + 4 \cdot 1}\right) - 31.2}{2}$$

$$\sigma_{ti} \approx -0.5\ MPa$$

2. Using the Mohr-Coulomb failure criterion (Figure 8.9), the unconfined compressive strength can be estimated as

$$\sigma_{ci} = 31.67 MPa$$

Solving the equation below

$$tan^2\left(45 + \frac{\phi}{2}\right) = 4.56$$

the friction angle is found as

$$\phi \approx 39.8^{\circ}$$

Using Equation 8.5, the cohesion will be estimated as

$$2c\tan\left(45+\frac{39.8}{2}\right)=31.67$$

$$c \approx 7.4\ MPa$$

8.7 Review quiz

1. The average vertical stress (σ_v) at a depth of z (m) can be estimated in MPa as

   a) $\sigma_v \approx 0.027z$ b) $\sigma_v \approx 2.7z$ c) $\sigma_v \approx 0.27z$ d) cannot be estimated

2. In rock mass, the horizontal stress is generally greater than the vertical stress at shallow depths.

   a) True b) False

3. For the Hoek-Brown failure criterion, the parameter s for intact rocks is equal to

   a) 0 b) 0.5 c) 0.616 d) 1

4. The Griffith strength criterion gives a good indication of

   a) compressive strength b) peak strength
   c) tensile strength d) shear strength

5. What criterion will provide a better estimation of the tensile strength of rock?

   a) Hoek-Brown b) Barton c) Mohr-Coulomb d) all three (a–c)

6. Jointed rocks with a higher JRC are typically associated with higher shear strength.

   a) True b) False

7. Barton's comb is used to measure

   a) RQD b) JRC c) JCS d) UCS

Answers: 1. a 2. a 3. d 4. c 5. a 6. a 7. b

Chapter 9

Rock mass ratings and properties

Project relevance: Engineering practice indicates that the properties of small-size rock specimens tested in the lab can be very different from the properties of the large rock mass which these specimens were collected from. The main reason for this is that rock mass often contains a set of discontinuities that affect its strength characteristics, while laboratory specimens are generally intact. This chapter will discuss how to estimate the rock mass properties using data from lab testing and site investigation. It will also explain how to obtain the strength parameters of the rock mass from the project site.

9.1 General considerations

Engineers have excavated rocks for mining and tunnel purposes for hundreds of years. Unfortunately, there have been several occasions where a section or the whole tunnel collapsed in weak or heavily weathered rocks. To avoid failure, engineers have developed rock mass classification schemes so that appropriate support measures can be readily determined and implemented. These classification systems often refer to 'active span' and 'stand-up time' of tunnels. Active span (S) is the largest dimension of the unsupported tunnel section (Figure 9.1). Stand-up time is the length of time which an excavated opening with a given active span can stand without any means of support or reinforcement.

Question: *What are the most commonly used classification systems?*
Answer: The rock mass rating (RMR) system and rock tunneling quality index (Q-system). The geological strength index (GSI) combined with the Hoek-Brown failure criterion is commonly used to determine the strength characteristics of rock mass.

9.2 Rock mass rating

Bieniawski (1973, 1989) analyzed data from a number of case records and developed the Geomechanics Classification of Rock Mass Rating (RMR) system to determine the quality of rock mass (Table 9.1). This system uses six major parameters of rock mass such as (1) strength of intact rock material; (2) rock quality designation (RQD); (3) spacing of joints;

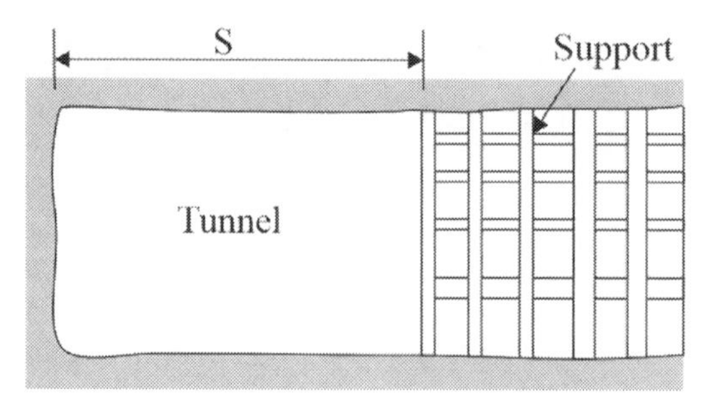

Figure 9.1 Definition of the active span (S)

Table 9.1 Rock mass rating (RMR)

Parameter								
1. Strength of intact rock	Point load index (MPa)	> 10	4–10	2–4	1–2			
	UCS (MPa)	> 250	100–250	50–100	25–50	5–25	1–5	< 1
Rating		15	12	7	4	2	1	0
2. RQD (%)	90–100	75–90	50–75	25–50	< 25			
Rating	20	17	13	8	3			
3. Joint spacing (m)	> 2	0.6–2	0.2–0.6	0.06–0.2	< 0.06			
Rating	20	15	10	8	5			
4. Condition of joints	Not continuous, very rough surfaces, unweathered, no separation	Slightly rough surfaces, slightly weathered, separation < 1 mm	Slightly rough surfaces, highly weathered, separation < 1 mm	Continuous, slickensided surfaces, or gouge < 5 mm thick, or separation 1–5 mm	Continuous, joints, soft gouge > 5 mm thick, or separation > 5 mm			
Rating	30	25	20	10	0			
5. Ground water	Inflow per 10 m tunnel length (L/min)	None	< 10	10–25	25–125	> 125		
	Joint water pressure/major in situ stress	0	0–0.1	0.1–0.2	0.2–0.5	> 0.5		
	General conditions at excavation surfaces	Dry	Damp	Wet	Dripping	Flowing		
Rating		15	10	7	4	0		

(4) condition of joints including aperture, persistence, roughness, joint surface weathering, alteration, and presence of filling; (5) ground water conditions; and (6) orientation of discontinuities. Tables 9.2 and 9.3 provide rating adjustments based on the direction of tunnel.

Question: *What is a slickensided surface, a term that is used to describe the joint conditions?*
Answer: Slickenside is a smoothly polished surface caused by frictional movements between rocks. The surface is normally striated in the direction of movement.

The weight of each parameter is considered in the rating, while the overall RMR is the sum of individual ratings (maximum 100 points). It is noted that sound judgment is often required to interpret the information provided by investigation reports.

Table 9.2 Rating adjustment for joint orientation

Effect of joints		*Very favorable*	*Favorable*	*Fair*	*Unfavorable*	*Very unfavorable*
Rating	Tunnels	0	−2	−5	−10	−12
	Foundations	0	−2	−7	−15	−25
	Slopes	0	−5	−25	−50	−60

Table 9.3 Effect of joint orientation in tunneling

Strike perpendicular to tunnel axis				*Strike parallel to tunnel axis*		*Dip 0°–20°*
Drive with dip		*Drive against dip*				
Dip 45°–90°	Dip 20°–45°	Dip 45°–90°	Dip 20°–45°	Dip 45°–90°	Dip 20°–45°	Irrespective of strike
Very favorable	Favorable	Fair	Unfavorable	Very unfavorable	Fair	Fair

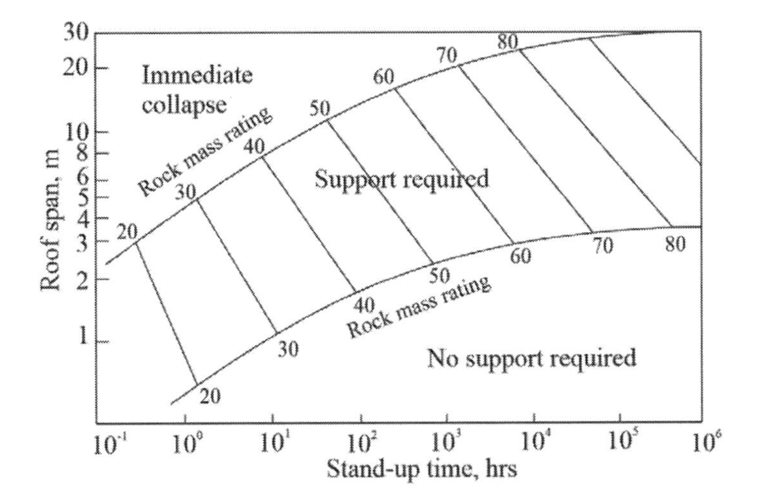

Figure 9.2 Relationship between the stand-up time and active span of tunnel

Question: *What is the practical application of this classification system?*
Answer: Based on the value of RMR, an appropriate support for tunnels can be selected. Figure 9.2 gives estimates for whether support is required or not. Detailed guidelines for tunnel support required for rock mass with different RMR values are given in Bieniawski (1989).

Question: *How do we use the tunnel adjustment for RMR?*
Answer: Adjustments are made according to the tunnel direction and orientation of bedding planes as given in Tables 9.2 and 9.3, and schematically shown in Figure 9.3.

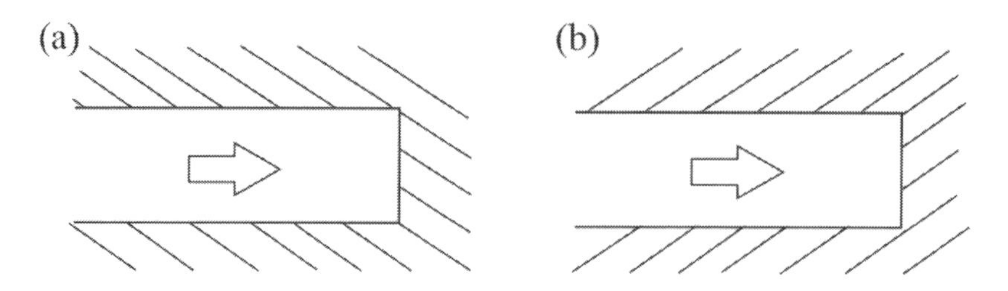

Figure 9.3 Adjustments are made according to the way a tunnel is constructed: (a) drive with dip and (b) drive against dip

Table 9.4 Classification of rock mass based on RMR values

RMR Rating	81–100	61–80	41–60	21–40	< 20
Rock mass class	A	B	C	D	E
Description	Very good rock	Good rock	Fair rock	Poor rock	Very poor rock

Rock mass with better rock quality is associated with a larger span and longer stand-up time. Class A is assigned to very good quality rock mass while Class E is assigned to very poor quality rock as shown in Table 9.4.

Question: *Is RMR used for mining purposes?*
Answer: The mining industry considers RMR as somewhat conservative. For this reason, RMR has been modified to include the in situ and induced stresses, stress changes and the effects of blasting and weathering, resulting in the modified rock mass rating (MRMR) (Laubscher and Page, 1990).

9.3 Rock tunnel quality Q-system

The Q-system was developed as a rock tunneling quality index by the Norwegian Geotechnical Institute (NGI) (Barton et al., 1974). The objective was to characterize the rock mass and determine the tunnel support requirements. Similar to RMR, the Q-system is based on numerous case studies related to underground excavation. The Q rating is obtained using Equation 9.1.

$$Q = \frac{RQD}{J_n} \cdot \frac{J_r}{J_a} \cdot \frac{J_w}{SRF} \tag{9.1}$$

where RQD measures the fracturing degree, J_n is the joint set number showing the number of joint sets, J_r is the joint roughness number accounting for the joint surface roughness, J_a is the joint alteration number indicating the degree of weathering, alteration and filling, J_w is the joint water reduction factor accounting for the problem from ground water pressure, and SRF is the stress reduction factor indicating the influence of in situ stresses.

The Q value is generally considered as a function of three parameters which are measures of (a) block size (RQD/J_n), (b) inter-block shear strength (J_r/J_a) and (c) active stress (J_w/SRF).

Unlike RMR, the Q value can range from 0.001 to 1000+. The several orders of magnitude range of Q is partially due to the large variability of geology and structural geology (Barton, 2007).

Table 9.5 RQD values for Q-system

Class	*Description*	*RQD*
A	Very poor	0–25
B	Poor	25–50
C	Fair	50–75
D	Good	75–90
E	Excellent	90–100

Note: When *RQD* < 10, use 10 in computing Q.

Table 9.6 Joint set number (J_n) for Q-system

Class	*Description*	*Joint set number*
A	Massive; none or few joints	0.5–1
B	One joint set	2
C	One joint set plus random joints	3
D	Two joint sets	4
E	Two joint sets plus random joints	6
F	Three joint sets	9
G	Three joint sets plus random joints	12
H	Four or more joint sets; heavily jointed	15
J	Crushed rock; earth like	20

Table 9.7 Joint roughness number (J_r) for Q-system (Barton et al., 1974)

Class	*Description*	*Joint roughness number*
(a) Rock-wall contact and (b) Rock-wall contact before 10 cm of shear		
A	Discontinuous joints	4
B	Rough or irregular, undulating	3
C	Smooth, undulating	2
D	Slickensided, undulating	1.5
E	Rough or irregular, planar	1.5
F	Smooth, planar	1
G	Slickensided, planar	0.5
(c) No rock-wall contact when sheared		
H	Zone containing clay minerals thick enough to prevent rock-wall contact	1
J	Sandy, gravelly or crushed zone thick enough to prevent rock-wall contact	1

Table 9.8 Joint alternation number (J_a) for Q-system (Barton et al., 1974)

Class	*Description*	φ_r (°)	*Joint alternation number*
(a) Rock-wall contact (no mineral fillings, only coatings)			
A	Tightly healed, hard, non-softening, impermeable filling (i.e., quarts or epidote)	–	0.75
B	Unaltered joint walls, surface staining only	25–35	1
C	Slightly altered joint walls, non-softening mineral coatings, sandy particles, clay-free disintegrated rock, etc.	25–30	2
D	Silty- or sandy-clay coatings, small clay fraction (non-softening)	20–25	3
E	Softening or low friction clay mineral coatings (i.e., kaolinite or mica). Also chlorite, talc, gypsum, graphite, etc., and small quantities of swelling clays	8–16	4
(b) Rock-wall contact before 10 cm shear (thin mineral fillings)			
F	Sandy particles, clay-free disintegrated rock, etc.	25–30	4
G	Strongly over-consolidated non-softening clay material fillings (continuous but < 5 mm thickness)	16–24	6
H	Medium or low over-consolidated softening clay material fillings (continuous but < 5 mm thickness)	12–16	8
J	Swelling clay fillings (i.e. montmorillonite) (continuous but < 5mm thickness). Value of J_a depends on percentage of swelling clay-size particles and access to water.	6–12	8–12
(c) No rock-wall contact when sheared (thick mineral fillings)			
K, L, M	Zones or bands of disintegrated crushed rock and clay (see G, H and J for description of clay conditions)	6–24	6, 8 or 8–12
N	Zones or bands of silty- or sandy-clay, small clay fraction	–	5
O, P, R	Thick continuous zones or bands of clay (see G, H and J for description of clay conditions)	6–24	10, 13 or 13–20

Table 9.9 Joint water reduction factor (J_w) for Q-system (Barton et al., 1974)

Class	*Description*	*Water pressure (kPa)*	*Joint water reduction*
A	Dry excavation or minor inflow (i.e. < 5 L/min locally)	< 100	1
B	Medium inflow or pressure, occasional outwash of joint fillings	100–250	0.66
C	Large inflow or high pressure in competent rock with unfilled joints	250–1000	0.5
D	Large inflow or high pressure, considerable outwash of joint fillings	250–1000	0.33
E	Exceptionally high inflow or water pressure at blasting, decaying with time	>1000	0.2–0.1
F	Exceptionally high inflow or water pressure, continuing without noticeable decay	>1000	0.1–0.05

Note: C to F are crude estimates. Special problems caused by ice formation are not considered.

Table 9.10a Stress reduction factor (SRF) for Q-system (Barton et al., 1974)

Class	*Description*	*Stress reduction factor (SRF)*
A	Multiple occurrences of weakness zones containing clay or chemically disintegrated rock, very loose surrounding rock (any depth)	10
B	Single-weakness zone containing clay or chemically disintegrated rock (depth of excavation ≤ 50 m)	5
C	Single-weakness zone containing clay or chemically disintegrated rock (depth of excavation > 50 m)	2.5
D	Multiple shear zones in competent rock (clay-free) (depth of excavation ≤ 50 m)	7.5
E	Single shear zone in competent rock (clay-free) (depth of excavation ≤ 50 m)	5
F	Single shear zone in competent rock (clay-free) (depth of excavation > 50 m)	2.5
G	Loose, open joints, heavily jointed (any depth)	5

Table 9.10b Stress reduction factor (SRF) for Q-system (Barton et al., 1974)

Class	*Description*	σ_c/σ_1	σ_θ/σ_1	*Stress reduction factor (SRF)*
(b) competent rock, rock stress problems				
H	Low stress, near surface, open joints	> 200	< 0.01	2.5
J	Medium stress, favorable stress condition	200–10	0.01–0.3	1
K	High stress, very tight structure. Usually favorable to stability; may be unfavorable to wall stability	10–5	0.3–0.4	0.5–2
L	Moderate slabbing after > 1 hour in massive rock	5–3	0.5–0.65	5–50
M	Slabbing and rock burst after a few minutes in massive rock	3–2	0.65–1	50–200
N	Heavily rock burst (strain-burst) and dynamic deformations in massive rock	< 2	> 1	200–400
(c) squeezing rock: plastic flow of incompetent rock under the influence of high rock pressure				
O	Mild squeezing rock pressure		1–5	5–20
P	Heavy squeezing rock pressure		> 5	10–20
(d) swelling rock: chemical swelling activity depending on presence of water				
R	Mild swelling rock pressure			5–10
S	Heavy swelling rock pressure			10–15

Note: σ_c is unconfined compressive strength, σ_1 and σ_3 are major and minor principal stresses and σ_θ is maximum tangential stress (estimated from elastic theory). For strongly anisotropic stress fields (when measured), when σ_1/σ_3 = 5 to 10, reduce σ_c by 25%, and for σ_1/σ_3 > 10, reduce σ_c by 50%.

Table 9.11 Rock mass classification based on Q-system

Q-value	*Class*	*Rock mass quality*
400–1000	A	Exceptionally good
100–400	A	Extremely good
40–100	A	Very good
10–40	B	Good
4–10	C	Fair
1–4	D	Poor
0.1–1	E	Very poor
0.01–0.1	F	Extremely poor
0.001–0.01	G	Exceptionally poor

9.4 Geological Strength Index

Geological strength index (GSI) (Table 9.12) was introduced by Hoek (1994) to estimate the reduction in rock mass strength for different geological conditions. For good quality rock mass, the following relationship between GSI and RMR can be used

$$GSI \approx RMR - 5 \tag{9.2}$$

Question: *The RMR and Q-system are widely used in practice. Are there any correlations between these two ratings?*
Answer: Yes, there are two equations that link the RMR and Q-system. Equation 9.3 was proposed by Bieniawski (1989), while Equation 9.4 was given by Barton (1995).

$$RMR \approx 9 \ln Q + 44 \tag{9.3}$$

$$RMR \approx 15 \log Q + 50 \tag{9.4}$$

9.5 Project work: rock mass ratings

RMR for all three geological units (sandstone, mudstone and andesite) will be estimated using the data from the field and laboratory investigations. Tables 9.13–9.15 explain how the RMR was obtained for each rock mass.

9.6 Rock mass properties

The generalized Hoek-Brown criterion can be used to estimate the strength characteristics of rock mass, which is essential in the assessment of rock mass stability.

$$\sigma'_{1f} = \sigma'_{3f} + \sigma_{ci}\left(m_b \frac{\sigma'_{3f}}{\sigma_{ci}} + s\right)^{\alpha} \tag{9.5}$$

where m_b is derived from the value of the intact rock m_i as

$$m_b = m_i \cdot exp\left(\frac{GSI - 100}{28 - 14D}\right) \tag{9.6}$$

Table 9.12 Geological strength index (GSI) from Hoek and Marinos (2000)

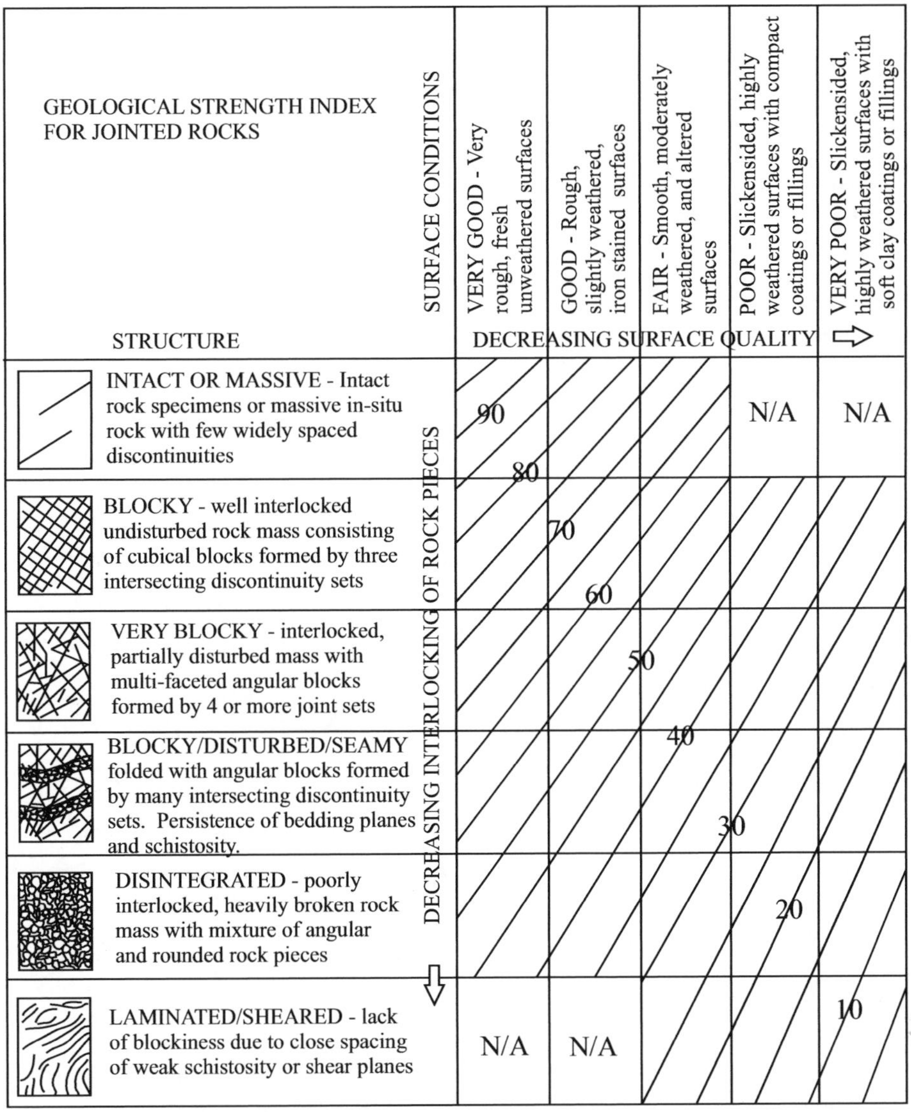

GEOLOGICAL STRENGTH INDEX FOR JOINTED ROCKS STRUCTURE | SURFACE CONDITIONS (DECREASING SURFACE QUALITY ⇨)	VERY GOOD - Very rough, fresh unweathered surfaces	GOOD - Rough, slightly weathered, iron stained surfaces	FAIR - Smooth, moderately weathered, and altered surfaces	POOR - Slickensided, highly weathered surfaces with compact coatings or fillings	VERY POOR - Slickensided, highly weathered surfaces with soft clay coatings or fillings
INTACT OR MASSIVE - Intact rock specimens or massive in-situ rock with few widely spaced discontinuities	90 80			N/A	N/A
BLOCKY - well interlocked undisturbed rock mass consisting of cubical blocks formed by three intersecting discontinuity sets		70 60			
VERY BLOCKY - interlocked, partially disturbed mass with multi-faceted angular blocks formed by 4 or more joint sets			50		
BLOCKY/DISTURBED/SEAMY folded with angular blocks formed by many intersecting discontinuity sets. Persistence of bedding planes and schistosity.			40	30	
DISINTEGRATED - poorly interlocked, heavily broken rock mass with mixture of angular and rounded rock pieces				20	
LAMINATED/SHEARED - lack of blockiness due to close spacing of weak schistosity or shear planes	N/A	N/A			10

DECREASING INTERLOCKING OF ROCK PIECES ⇩

Table 9.13 Rock mass rating for the sandstone rock mass

Category	*Explanation*	*Rating*
1. Strength	UCS = 22.1 MPa (triaxial data analysis – Section 8.3)	2
2. RQD	RQD ≈ 26% (borehole data analysis – Section 5.5)	8
3. Joint spacing	From the relationship between RQD and joint spacing given in Figure 6.4, the joint spacing is estimated as 0.09–0.1 m	8
4. Conditions of joints	Joints are slickensided, undulating, highly weathered. Joints are separated by about 3–5 mm and filled with clay (data from borehole logs – Section 2.1).	10
5. Ground water	No information regarding the ground water level, the rock was "moist" during the borehole exploration (data from borehole logs – Section 2.1)	10
Total rating		38

Table 9.14 Rock mass rating for the mudstone rock mass

Category	*Explanation*	*Rating*
1. Strength	UCS = 44.8 MPa (point load data analysis – Section 7.2.8)	4
2. RQD	RQD ≈ 77.5% (borehole data analysis – Section 5.5)	17
3. Joint spacing	From the relationship between RQD and joint spacing given in Figure 6.4, the joint spacing is estimated as 0.25 m	10
4. Conditions of joints	Joints surfaces are slightly rough, slightly weathered with stains, no clay found on surface, apertures generally less than 1 mm (data from borehole logs – Section 2.1)	25
5. Ground water	No information regarding the ground water level, the rock seems to be dry (data from borehole logs – Section 2.1)	15
Total rating		71

Table 9.15 Rock mass rating for the andesite rock mass

Rock mass parameters	*Description*	*Rating*
1. Strength	UCS = 60.0 MPa (UCS test analysis – Section 7.2.6)	7
2. RQD	RQD ≈ 92% (borehole data analysis – Section 5.5)	20
3. Joint spacing	From the relationship between RQD and joint spacing given in Figure 6.4, the joint spacing is estimated as 0.4–0.5 m	10
4. Conditions of joints	Joints surfaces are generally stepped and rough, tightly closed and unweathered with occasional stains (data from borehole logs – Section 2.1)	30
5. Ground water	No information regarding the ground water level, the rock seems to be dry (data from borehole logs – Section 2.1)	15
Total rating		82

where D is the disturbance factor which accounts for the disturbance in the rock mass due to blasting or mechanical excavation. It varies from 0 to 1, with 0 for undisturbed and 1 for highly disturbed rock mass. The constant s and α can be estimated as follows:

$$s = exp\left(\frac{GSI-100}{9-3D}\right) \tag{9.7}$$

$$\alpha = \frac{1}{2} + \frac{1}{6}\left(e^{-GSI/15} - e^{-20/3}\right) \tag{9.8}$$

The constant s generally varies from 0 to 1, with 0 for poor quality rock and 1 for intact rock. The constant α varies between 0.5 for good quality rock and 0.65 for poor quality rock.

The strength of rock mass can be estimated as

$$\sigma_{cm} = \sigma_{ci} \cdot s^{\alpha} \tag{9.9}$$

where σ_{ci} is the UCS of the intact rock.

To estimate the friction angle and cohesion of the rock mass, Hoek and Brown (1997) proposed two charts (Figures 9.4 and 9.5) that provide correlations between the m_i parameter, GSI and rock mass strength.

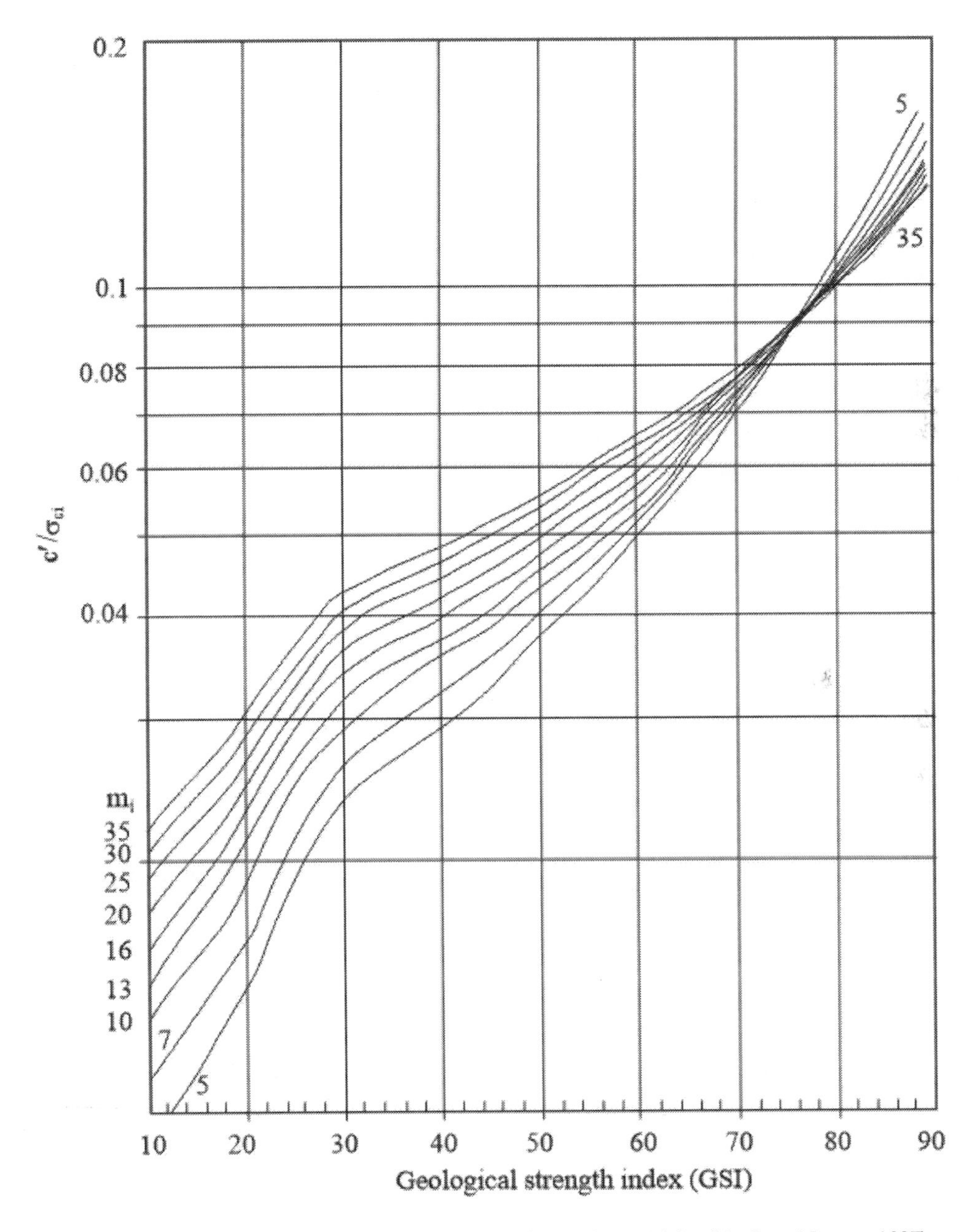

Figure 9.4 Relationship between cohesion, m_i and GSI for rock mass (after Hoek and Brown, 1997)

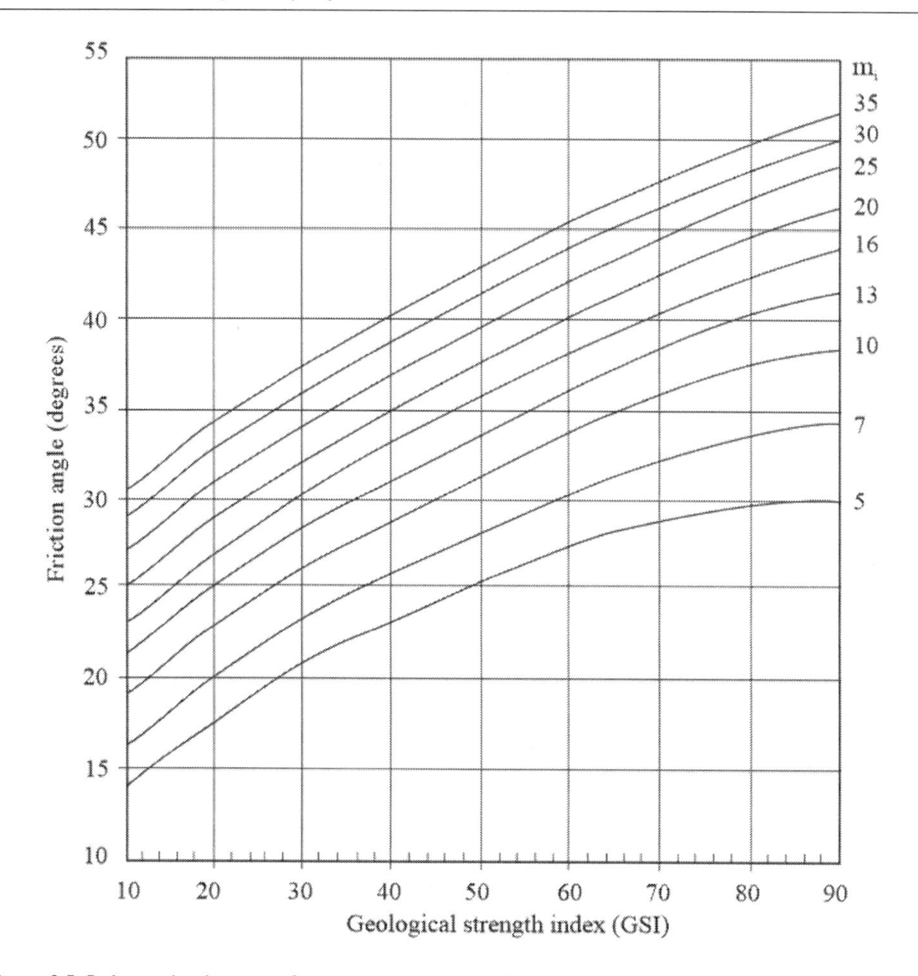

Figure 9.5 Relationship between friction angle, m_i and GSI for rock mass (after Hoek and Brown, 1997)

9.7 Project work: rock mass properties

The UCS of rock mass (σ_{cm}) for sandstone, mudstone and andesite can be determined as follows. The example below shows how to estimate the rock mass properties for the sandstone layer

$$\sigma_{ci} = 22.1\ MPa$$

$$GSI \approx RMR - 5 = 38 - 5 = 33$$

$$s = exp\left(\frac{GSI - 100}{9 - 3D}\right) = exp\left(\frac{33 - 100}{9 - 3 \cdot 0}\right) \approx 0.00058$$

$$\alpha = \frac{1}{2} + \frac{1}{6}\left(e^{-GSI/15} - e^{-20/3}\right) = \frac{1}{2} + \frac{1}{6}\left(e^{-33/15} - e^{-20/3}\right) = 0.518$$

$$\sigma_{cm} = \sigma_{ci} \cdot s^{\alpha} = 22.1 \cdot 0.00058^{0.518} \approx 0.47\ MPa$$

Table 9.16 Hoek-Brown constants and strength characteristics of three rock masses

	Sandstone	*Mudstone*	*Andesite*
σ_{ci} (MPa)	22.1	44.8	60
GSI = RMR - 5	33	66	77
s	0.00058	0.023	0.078
α	0.518	0.502	0.501
σ_{cm} (MPa)	0.47	6.73	16.69
m_i	15.2	6	25
c'/σ_{ci}	0.037	0.06	0.09
c' (MPa)	0.82	2.69	5.4
φ (°)	31	31	46

Table 9.16 summarizes the Hoek-Brown constants and σ_{cm} values for all three geological units.

To estimate the friction angle and cohesion, Figures 9.4 and 9.5 will be used. For the sandstone rock mass with $m_i = 15.2$ and GSI = 33 (Figure 9.4), the ratio between the cohesion (c) and the UCS of intact rock (σ_{ci}) is about 0.037. The cohesion of rock mass equals

$$c' = 0.037 \cdot \sigma_{ci} = 0.037 \cdot 22.1 \approx 0.82\ MPa$$

From Figure 9.5, the friction angle for the sandstone rock mass is estimated as 31°.

The cohesion and friction angle for the rock mass of mudstone and andesite are given in Table 9.16.

Question: *The m_i parameter (m_i = 15.2) for the sandstone has been obtained from triaxial tests. How do you obtain this m_i parameter for the mudstone and andesite?*

Answer: We obtain these values from Table 9.17, which summarizes m_i based on the rock origin (type). For andesite it is 25; however, there is no m_i value given for mudstone. For this reason, we will assume it to be around 6, the value for shale, as these two rocks have similar geologic origin.

9.8 Problems for practice

Problem 9.1: Estimate the water pressure (kg/cm^2) in a tunnel excavated in jointed sandstone (UCS = 50 MPa). The tunnel will be at 70 m below the ground level and the ground water table is 10 m below the ground surface.

Solution:

At 70 m, the water depth is 60 m (70 − 10=60), which is equivalent to 6 kg/cm^2.

Table 9.17 m_i values for different types of rock (after Hoek and Brown, 1997)

Rock type	*Texture*			
Sedimentary	Coarse	Medium	Fine	Very fine
	Conglomerate (21 ± 3)	Sandstone (17 ± 4)	Siltstone (7 ± 2)	Claystone (4 ± 2)
	Breccia (19 ± 5)		Greywacke (18 ± 3)	Shale (6 ± 2) Marl (7 ± 2)
	Crystalline limestone (12 ± 3)	Sparitic limestone (10 ± 2) Gypsum (8 ± 2)	Micritic limestone (9 ± 2) Anhydrite (12 ± 2)	Dolomite (9 ± 3) Chalk (7 ± 2)
Metamorphic	Marble (9 ± 3)	Hornfels (19 ± 4) Metasandstone (19 ± 3)	Quartzite (20 ± 3)	
	Migamatite (29 ± 3)	Amphibiolite (26 ± 6) Schist (12 ± 3)	Gneiss (28 ± 5) Phyllite (7 ± 3)	Slate (7 ± 4)
Igneous	Granite (32 ± 3)	Diorite (25 ± 5)		
	Gabbro (27 ± 3)	Dolerite (16 ± 5)		
	Porphyry (20 ± 5)		Diabase (15 ± 5)	Peridotite (25 ± 5)
		Rhyolite (25 ± 5)	Basalt (25 ± 5)	Obsidian (19 ± 3)
		Andesite (25 ± 5)		
	Agglomerate (19 ± 3)		Tuff (13 ± 5)	

Problem 9.2: A tunnel will be excavated in sandstone rock mass (density is 2.3 g/cm^3) at a depth of 70 m. At this depth, the sandstone has bedding planes with an orientation of 090/40, creating a set of joints which may be critical to the stability of the tunnel. Joints in this set are highly weathered with slightly rough surfaces. The strength of the intact rock is 75 MPa, and values for the RQD are reported as 70%. The average joint frequency is 8 joints/m. The ground water existed at the tunnel depth, with the ground water table was found at 15 m below the ground surface. The tunnel is being driven from west to east. Estimate the RMR value including the rating adjustment for tunneling.

Solution:

Referring to Tables 9.1–9.3, RMR can be estimated as follows:

1. UCS is 75 MPa. Rating is 7.
2. RQD = 70%. Rating is 13.
3. 8 joints per m is equal to spacing of 1/8 = 0.125 m. Rating is 8.
4. Highly weathered and slightly rough surfaces. Rating is about 20.
5. The ratio between the pore water pressure (u) and normal stress (σ) at a depth of the tunnel (70 m) will be used. The pore water pressure equals

$$u = 9.81 \cdot (70 - 15) \approx 539 kPa$$

The normal stress is

$$\sigma = \gamma \cdot z = 2.3 \cdot 9.81 \cdot 70 \approx 1577.8 \ kPa$$

The u/σ ratio is =539/1577=0.34. Rating is 4.

5. Tunnel adjustment is 'drive with dip'. A dip of 40° makes it 'favorable'. Rating is −2.

$$RMR = 7 + 13 + 8 + 20 + 4 - 2 = 50$$

Problem 9.3: A sandstone rock mass is fractured by three joint sets plus random fractures. The average RQD is 75%; the average joint spacing is 0.18 m. The joint surfaces are slightly rough and slightly weathered. The joints are in contact with apertures generally less than 1 mm; no clay is found on the surfaces. The point load strength index of the sandstone is 3.5 MPa. The tunnel is to be excavated at 50 m below the ground level and the ground water table is 20 m below the ground surface. Estimate the *Q*-value.

Solution:

Referring to Tables 9.5–9.10, the *Q* value can be obtained as follows:

1. RQD = 75.
2. $J_n = 12$ (three joints plus random fractures).
3. $J_r = 1.5$ (slightly rough and slightly weathered).
4. $J_a = 2$ (no clay found on the surfaces).
5. $J_w = 0.5$ (water pressure results from 30 m of water over the tunnel (50 − 20=30 m), which is equivalent to 3 kg/cm^2).
6. SRF is obtained from the ratio between rock's UCS and the normal stress (σ) at a depth of the tunnel.

$$UCS = 24 \cdot PLI = 24 \cdot 3.5 = 84 \ MPa$$

$$\sigma \approx 0.027 \cdot z = 0.027 \cdot 50 = 1.35 \ MPa$$

The ratio UCS/σ is 84/1.35=62.2. Rating is 1.
Q rating is (Equation 9.1)

$$Q = \frac{75}{12} \cdot \frac{1.5}{2} \cdot \frac{0.5}{1} \approx 2.3$$

The rock mass is of poor quality (Table 9.11).

9.9 Review quiz

1. If RMR = 50, what is the maximum span (in meters) of a tunnel that doesn't require support (see Figure 9.2)?

a) 1.0 b) 1.3 c) 1.7 d) 2.2

2. In tunneling, the active span is referred to

   a) the largest part of the tunnel section with support
   b) the largest diameter of the tunnel
   c) the orientation of the tunnel towards the active principal stress
   d) the largest dimension of the tunnel section without support

3. What parameter is NOT used in the rock mass rating (RMR)?

   a) UCS b) RQD c) JRC d) ground water

4. Using RMR, the geological strength index (GSI) can be estimated as follows:

   a) GSI ≈ RMR b) GSI ≈ RMR − 10
   c) GSI ≈ RMR − 5 d) GSI ≈ RMR/10

5. For rock mass with a massive structure, the GSI varies in the range of

   a) 20–40 b) 40–60 c) 60–80 d) 80–100

6. A tunnel is being excavated from north to south in rock mass with bedding planes. The joints in these bedding planes have an orientation of 180/55. Select the correct statement.

   a) the excavation 'drives with dip' b) the excavation 'drives against dip'
   c) the excavation 'drives along the strike' d) none of the statements (a–c) is correct

7. Q-value is considered to be a function of three parameters which can be described as (1) inter-block shear strength, (2) active stress, and

   a) block size b) pore water pressure
   c) tunnel characteristics d) degree of weathering

8. What rock, when fresh, will generally have a higher m_i?

   a) granite b) sandstone c) shale d) siltstone

Answers: 1. d 2. d 3. c 4. c 5. d 6. a 7. a 8. a

Chapter 10

Rock falls

Project relevance: Rock falls commonly occur on steep slopes and cause significant economic damage. This chapter will discuss the factors leading to rock falls, outline methods of hazard assessment and discuss commonly used approaches to minimize the damage from this disaster. As part of the project work, rock fall hazard assessment will be conducted and different methods to prevent or mitigate the impact of this natural phenomenon will be discussed.

10.1 Rock falls and factors affecting them

Rock fall is a natural disaster that causes significant economic loss to local communities through blocking transportation routes and slowing traffic (Figure 10.1). According to Rapp (1960), rock falls can be classified based on the source of detachment as the primary rock fall (when bedrocks are involved) and the secondary rock fall (when the rock fall consists of rock fragments from primary rock falls).

Factors that cause rock falls (Table 10.1) include slope excavation, root growth, water pressure in joints, ground shaking, loss of support due to erosion and chemical weathering.

10.2 Rock fall characteristics

There are three major types of rock motion (Figure 10.2) in respect to the slope inclination:

1. Free fall (slope angle from 80° to 90°)
2. Bouncing (slope angle from 45° to 80°)
3. Rolling (slope angle less than 45°)

To predict the rock trajectory, it is necessary to know the properties of falling rocks (their size and shape), slope geometry (e.g., the slope height and inclination) and the coefficient of restitution (CoR). CoR can be obtained through a series of lab tests with a falling rock and rock surface (Gratchev and Saeidi, 2019) as schematically shown in Figure 10.3.

Figure 10.1 Rock fall that originated in the mountainous area (background of this image) during earthquake and transformed to a rock avalanche, resulting in casualties

Table 10.1 Major rock fall causes on highways in California (after Wyllie and Mah, 2014)

Causes of rock falls	*Percentage of falls*
Rain	30
Freeze-thaw	21
Fractured rocks	12
Wind	12
Snowmelt	8
Channeled runoff	7
Adverse planar fracture	5
Burrowing animals	2
Differential erosion	1

In such tests, the velocity of rock before (V_{before}) the contact with the rock surface and after (V_{after}) is measured and used to calculate CoR (Equation 10.1).

$$CoR = \frac{V_{after}}{V_{before}} \tag{10.1}$$

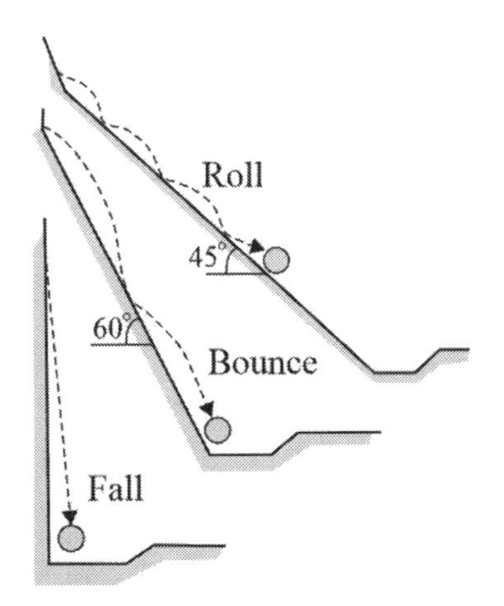

Figure 10.2 Falling rock motion

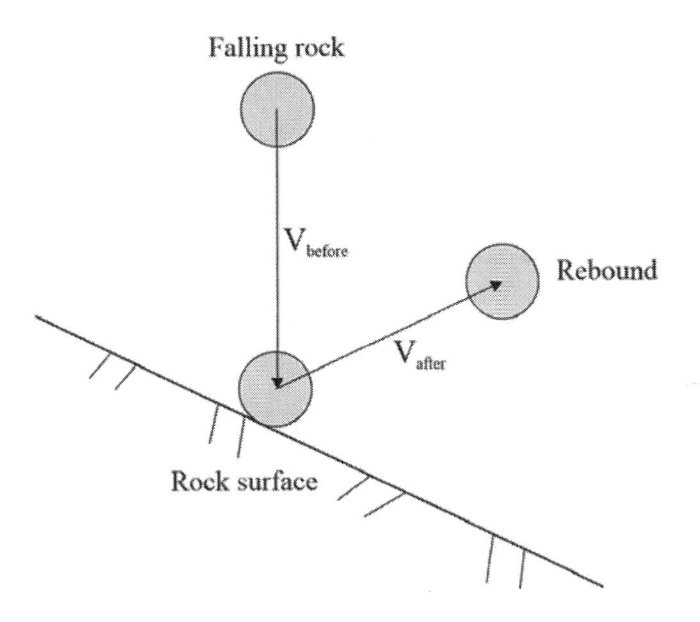

Figure 10.3 Definition of coefficient of restitution (CoR)

Results of previous field and lab studies indicate that CoR typically ranges from 0.4 to 0.8.

10.3 Rock fall hazard assessment

Rock fall hazard assessment is a complex process which aims at identifying sites with a high probability of rock fall occurrence. It includes regular site visits, rock fall modeling and estimation of damage. Figure 10.4 shows three zones related to a rock fall event: (1)

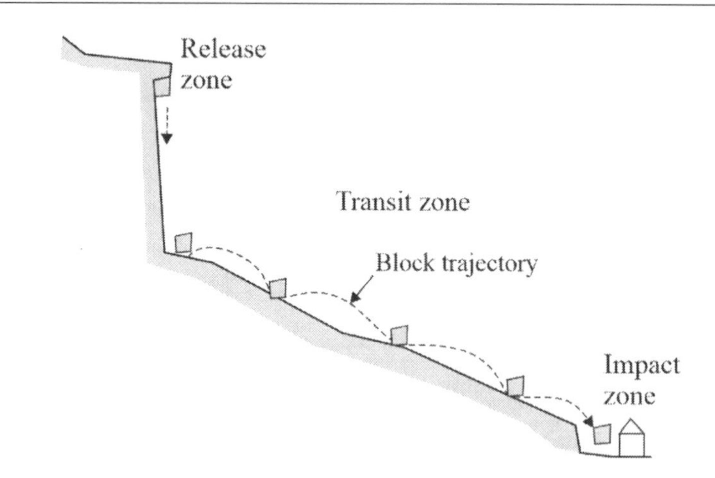

Figure 10.4 Assessment of rock fall hazard: three zones related to rock fall event

release zone, where the rock fall initiates; (b) transit zone, where the falling rock moves downhill; and (c) impact zone, where the rock may come in contact with engineering structures.

Regular site surveys help to identify slopes with a high probability of rock fall. During such visits, rock fall characteristics of each slope are assessed, and a rock fall hazard rating is assigned, while sites with relatively high ratings are selected for more detailed investigation. An example of a rock fall hazard rating system used in North America for major transportation routes is given in Table 10.2. Rock fall factors related to slope geometry, geology and roadside conditions are considered and rated accordingly. The final score, which is the summation of rating points, determines whether the site needs immediate remediation measures (a rating of 500 or more) or it can be considered as low priority (a rating of 300 or less) and thus no urgent measures are required.

Note that

$$AVR = \frac{ADT \cdot L}{S} \cdot 100\% \tag{10.2}$$

where AVR is the average vehicle risk, ADT is the average daily traffic (cars/hour), L is the slope length (km) and S is the posted speed limit (km/h).

$$DSD = \frac{\text{Actual site distance}}{\text{Decision site distance}} \cdot 100\% \tag{10.3}$$

where DSD is the decision site distance, which is related to the posted speed limit as given in Table 10.3.

10.4 Project work: rock fall hazard assessment

The rock fall hazard at a site near the dam (Point R in Figure 2.1) where a local road (the road width is 9.5 m) is frequently blocked by rocks during a rainfall season will be assessed. The average annual rainfall is high, ranging from 1300 to 3000 millimeters. The dominant geology of the 20 m high slope is sandstone with well-defined bedding

Table 10.2 Rock Fall Hazard Rating system (after Pierson, 1992)

Points	*3*	*9*	*27*	*81*
1. Slope height (m)	7.5	15	23	30
2. Ditch effectiveness	Good catchment	Moderate catchment	Limited catchment	No catchment
3. Average vehicle risk (% of time)	25	50	75	100
4. Percentage of decision sight distance (% of design value)	Adequate sight distance 100	Moderate sight distance 80	Limited sight distance 60	Very limited sight distance 40
5. Roadway width including paved shoulders (m)	13.5	11	8.5	6
6. Geologic character Case 1: Structural conditions Rock friction	Discontinuous joints, favorable orientation Rough, irregular	Discontinuous joints, random orientation Undulating	Discontinuous joints, adverse orientation Planar	Discontinuous joints, adverse orientation Clay infilling or slickensided
Case 2: Structural conditions Difference in erosion rates	Few differential erosion features Small difference	Occasional erosion features Moderate difference	Many erosion features Large difference	Major erosion features Extreme difference
7. Block size (m) or Volume of rock fall event (m^3)	0.3 3	0.6 6	1 9	1.2 12
8. Climate and presence of water on slope	Low to moderate precipitation; no freezing periods, no water on slope	Moderate precipitation; or short freezing periods, or intermittent water on slope	High precipitation or long freezing periods or continual water on slope	High precipitation and long freezing periods or continual water on slope and long freezing periods
9. Rock fall history	Few falls	Occasional falls	Many falls	Constant falls

Table 10.3 Relationship between the posted speed limit and decision site distance

Posted speed limit (km/h (mph))	*Decision site distance (m (ft))*
48 (30)	137 (450)
64 (40)	183 (600)
80 (50)	229 (750)
97 (60)	305 (1000)
113 (70)	335 (1100)

planes. There are a number of joints with an adverse orientation. The joints typically reach a length of 4–5 m. The block size of falling rocks ranges from 0.2 m to 0.4 m. The surface of joints is smooth with no undulations. The traffic along the hazardous

Table 10.4 Rock fall hazard assessment for the site marked R in Figure 2.1

Category	Rating
1. Slope height (m)	18
2. Ditch effectiveness	27
3. AVR (%)	3
4. DSD (%)	73
5. Roadway width (+ paved shoulders)	17
6. Geologic character	27
7. Block size/quantity of rock fall event	3
8. Climate and presence of water on slope	27
9. Rock fall history	27
Rating	222

part of the slope (the length is about 500 m) is slow – about 30 cars per hour. Cars typically move at the speed limit of 60 km/h, and the site distance is about 77 m.

The rock fall hazard assessment of this slope is detailed in Table 10.4.

Note that

$$AVR = \frac{30 \cdot 0.5}{60} \cdot 100\% = 25\%$$

$$DSD = \frac{77}{171} \cdot 100\% = 45\%$$

The rock fall hazard rating is 222, which makes it the low priority slope, suggesting that no immediate remediation measures are needed. However, regular site inspections should be performed to monitor possible changes in the slope conditions that can affect the rating.

10.5 Rock fall protection

There are a few rock fall protection methods, which are used to minimize the damage from this natural disaster (Kim et al., 2015). These measures include (a) 'scaling' to remove loose rocks from the face of the slope, (b) benches (or berms) to absorb the energy of falling rocks (Figure 10.5), (c) ditches to collect falling rocks at the toe of the slope, (d) fences (Figure 10.6) or walls to prevent falling rocks from reaching engineering structures, (e) anchored hanging mesh (Gratchev et al., 2015) (Figure 10.7) and (f) rock sheds (Figure 10.8) to protect mountain roads.

Table 10.5 gives the energy absorption capacity (EAC) of commonly used protection systems. The higher EAC is related to systems that provide more resistance to falling rocks (or can catch larger rock blocks); however, it also increases the overall cost.

10.6 Project work: rock fall protection

To prevent the local road from falling rocks near the dam (Point R in Figure 2.1), a ditch option will be considered. Figure 10.9a gives a schematic setup of the slope at Point R.

Figure 10.5 Bench used to catch falling rocks

Figure 10.6 Fence to protect road from falling rocks

Figure 10.7 Anchored hanging mesh

Figure 10.8 Rock sheds used to protect mountain roads

Table 10.5 Energy absorption capacity (in kJ) of protection systems (after Descoeudres et al., 1999)

Rock fall protection	*Energy absorption capacity (kJ)*
Rigid barriers	10–50
Low energy fences	30–220
Fences and cables	200–1500
Sheds (galleries)	200–2000
Embankments	1000–5000
Reinforced dams	3000–30,000

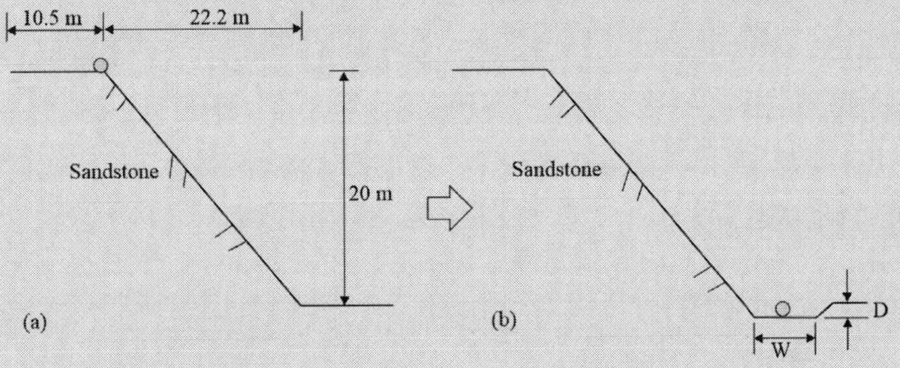

Figure 10.9 Slope geometry (a) and ditch parameters (b). W is the width, and D is the depth.

To design a ditch, knowledge of the slope's height and inclination is required. The ditch design chart is given in Figure 10.10.

The slope inclination is estimated from Figure 10.9. The slope height is 20 m and the slope angle equals

$$\tan\alpha = \frac{20}{22.2} \approx 0.9$$

where $\alpha \approx 42°$.

From the chart in Figure 10.10, a width of 4.57 m and a depth of 1.83 m can be obtained.

10.7 Problems for practice

Problem 10.1: About 120 cars pass this 35 m high slope every hour. The road site (600 m long) is often closed during rainy periods (about 3500 mm of precipitation per year) when many rock falls occur blocking the road (the road width is about 8.5 m). The size of the falling blocks typically reaches 1 m. The speed limit is 60 km/h, and the site distance is 60 m. There are a number of long joints on the slope with adverse orientation having smooth joint surfaces.

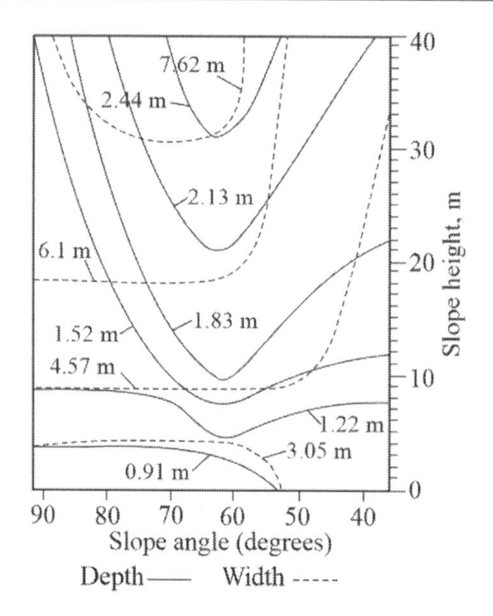

Figure 10.10 Ditch design chart for rock fall catchment (after Ritchie, 1963)

Table 10.6 Rock fall hazard assessment for the site described in Problem 10.1

Category	*Rating*
1. Slope height (m)	100
2. Ditch effectiveness	27
3. AVR (%)	81
4. DSD (%)	81
5. Roadway width (+ paved shoulders)	27
6. Geologic character	27
7. Block size/quantity of rock fall event	27
8. Climate and presence of water on slope	27
9. Rock fall history	27
Rating	424

Solution:

Each rock fall factor is assessed and rated as shown in Table 10.6

Note that

$$AVR = \frac{120 \cdot 0.6}{60} \cdot 100\% = 120\%$$

$$DSD = \frac{60}{171} \cdot 100\% = 35\%$$

Problem 10.2: Suggest rock fall protection measures for the slope given in Figure 10.11.

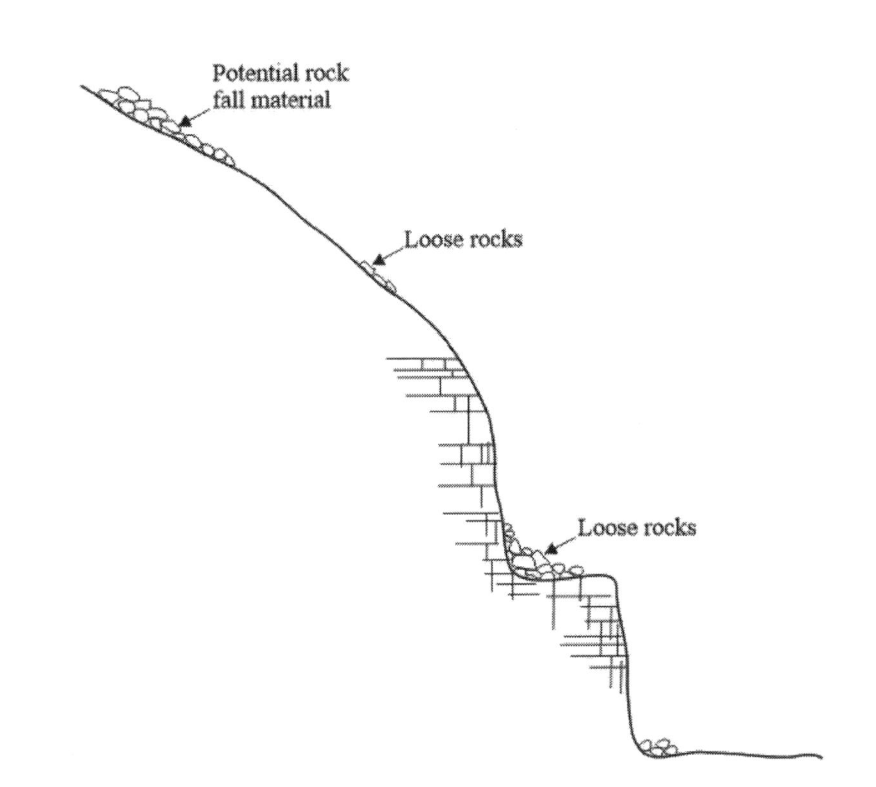

Figure 10.11 Slope with rock fall history

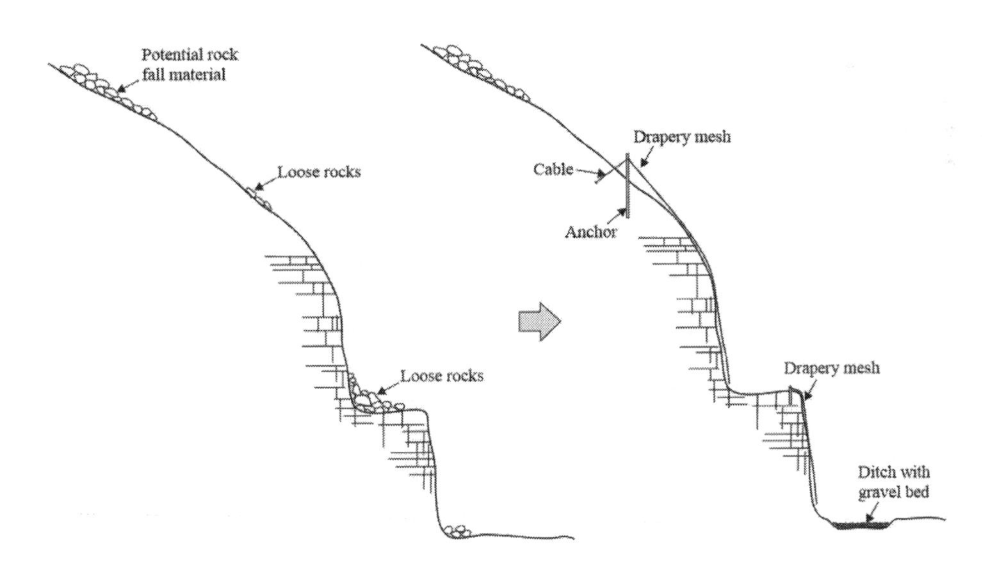

Figure 10.12 Slope protection against rock falls

Solution:

Figure 10.12 shows some measures that can prevent or minimize the effect of rock falls. Rock fall protection measures include (Figure 10.12)

1. Scaling of loose rocks on the slope;
2. Constructing a ditch at the toe of the slope (with a gravel bed);
3. Installing a drapery mesh system to minimize the impact of falling rocks;
4. The existing bench seems to work well, catching rocks from the top of the slope.

10.8 Review quiz

1. Wind can be a cause of rock falls.

   a) True b) False

2. When the overall slope angle is in the range of 60°–70°, what type of falling rock motion typically occurs?

   a) bounce b) free fall c) roll d) all three (a–c)

3. In rock fall hazard assessment, when falling rocks frequently reach the roadway, the ditch/catchment is described as

   a) no catchment b) limited c) moderate d) good

4. To design a ditch, engineers need to know

   a) slope height and falling rock weight
   b) slope angle and falling rock shape
   c) falling rock weight and their shape
   d) slope height and slope angle

5. To remove loose rocks from the slope, what method is used?

   a) scaling
   b) removing
   c) rock hiking
   d) rock displacing

6. Which of the following rock fall protection systems has the lowest energy absorption capacity?

   a) embankments b) fences + cable c) low energy fences d) rigid barriers

7. What will be a typical value of the coefficient of restitution for fresh greywacke?

   a) 0.04 b) 0.7 c) 1.5 d) 2.7

Answers: 1. a 2. a 3. b 4. d 5. a 6. d 7. b

Chapter 11

Rock slope stability

Project relevance: Natural slopes made of weathered rock material may experience stability issues during rainfall events. When saturated, such weathered rock typically loses some of its strength, which may result in slope movements. This chapter will introduce major landslide causes and triggers and outline the principles of slope stability analysis for rock mass. The slope stability at the project site with considerations of different ground water conditions will be analyzed and discussed.

11.1 Landslide triggers and causes

When engineers deal with landslides, they should understand the factors that can cause them and processes that can trigger them. The natural landslide triggers such as rainfall (or snowmelt), earthquake and volcano activity can set the landslide mass in motion. Landslide causes are factors that lead to the formation of the landslide mass. According to Varnes (1978). there are three types of landslides causes: geological, morphological and human (Table 11.1).

Rainfall and ground water conditions appear to be the major causes of landslides in rocks. The effect of these two factors on the slope stability is significant when the slope is made of porous and weathered rocks with high values of hydraulic conductivity (Figure 11.1). In addition, weathering can greatly decrease the strength of rocks as it can transform hard rock to soil-like material (Gratchev and Kim, 2016; Cogan et al., 2018). For example, Table 11.2 summarizes the properties of residual soil from Queensland (Australia), which was originated from different 'parent' rocks.

11.2 Types of slope failures

Depending on the type of motion, slope failures in rock mass can be classified as rock topple (Figure 11.2), rock slide (Figure 11.3) and rock spread (Figure 11.4) (Varnes, 1978).

Toppling (Figure 11.2) involves overturning of rock layers like a series of cantilever beams in slates, schists and bedded sediments inclined steeply into the hillside (Goodman, 1989). In such rock mass, each layer bends downhill under its own weight.

Most of rock slides tend to be translational where the rock mass moves along a roughly planar surface with little rotation or backward tilting. However, rotational slides can also

Table 11.1 Landslide causes (after Varnes, 1978)

Landslide causes	*Examples*
1. Geological causes	Weak or sensitive material
	Weathered material
	Sheared, jointed or fissured material
	Adversely oriented discontinuities (bedding, schistosity, fault)
2. Morphological causes	Tectonic of volcanic uplift
	Fluvial, wave or glacial erosion of slope toe
	Freeze-and-thaw weathering
3. Human causes	Excavation of slope or its toe
	Loading of slope or its crest
	Mining
	Deforestation

Figure 11.1 The landslide, which occurred in heavily weathered rhyolite (geological cause), was triggered by a 3-day rainfall event.

Table 11.2 Properties of residual soils from Queensland (Australia). The values show typical ranges of soil unit weight and strength (after Priddle et al., 2013)

Material property	*Parent rock type*		
	Mudstone	*Sandstone*	*Tuff*
Bulk density (g cm^{-3})	1.9–2.25	1.89–2.25	1.84–2.23
Cohesion (kN m^{-2})	62–360	21–269	24–176
Friction angle (°)	2.3–25	4–33	3.2–28.9

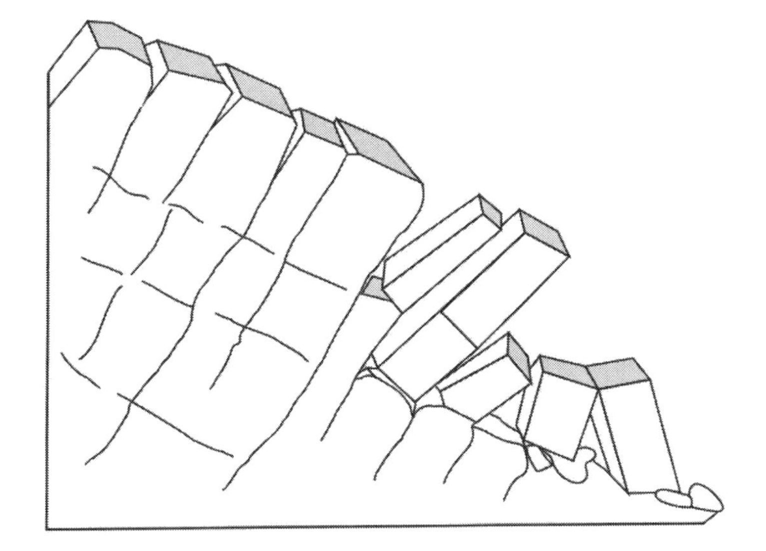

Figure 11.2 Rock toppling that commonly occurs along steeply inclined bedding planes

Figure 11.3 Large rock slide in weathered sandstone which was triggered by the 2004 Mid-Niigata Prefecture earthquake (Gratchev et al., 2006).

occur in heavily weathered rock mass or they can be triggered by a strong earthquake motion, as the one shown in Figure 11.3.

Lateral spreads (Figure 11.4) are very large landslides with a very gentle inclination of the failure plane (< 10°). The failure plane typically forms in a weak clay layer located underneath the massive rock.

Figure 11.4 Part of the Aratozawa landslide triggered by the 2008 Iwate-Miyagi Nairiku earthquake, Japan. This lateral spread failure in massive mudstone and siltstone rock mass had a length of 1200 m and a width of 800 m. The failure plane developed in a soft clay layer, which was inclined at 9°. More details can be found in Gratchev and Towhata (2010).

11.3 Slope stability analysis

Slope stability analysis is a procedure that engineers employ to assess the stability of natural and man-made slopes. It involves the use of the factor of safety (FS), which is defined as the ratio between the available and required strength (Equation 11.1):

$$FS = \frac{Available\ strength}{Required\ strength} \tag{11.1}$$

where the available strength is the maximum strength of soil mass as given in Equation 8.4 and the required strength is the strength that is necessary to keep the slope stable.

When $FS > 1$, the slope is considered to be stable; for $FS < 1$, the slope is assumed to be unstable; if $FS = 1$, the slope is under critical conditions.

11.3.1 Infinite slope

Many failures occur on slopes where heavily weathered rocks are underlain by hard bedrock. Such landslides are characterized by relatively shallow depths (1–2 m) and triggered by rainfall or earthquakes. The stability of such slopes can be analyzed using the infinite slope method as schematically shown in Figure 11.5.

The factor of safety for a dry slope (no ground water) can be estimated using Equation 11.2, while the effect of water seepage on slope stability can be evaluated using Equation 11.3.

$$FS = \frac{c}{\gamma h \sin\beta \cos\beta} + \frac{\tan\phi}{\tan\beta} \tag{11.2}$$

$$FS = \frac{c}{\gamma h \sin\beta \cos\beta} + \frac{\gamma - \gamma_w}{\gamma} \cdot \frac{\tan\phi}{\tan\beta} \tag{11.3}$$

Problem 11.1: The following slope and rock characteristics are considered: the thickness of the weathered rock layer is 2 m while the slope inclination is 30°. The unit weight of the weathered rock is 20 kN/m^3, cohesion is 10 kPa and the friction angle is 35°. Calculate the factor of safety when

1. The slope is dry
2. Seepage is parallel to the slope.

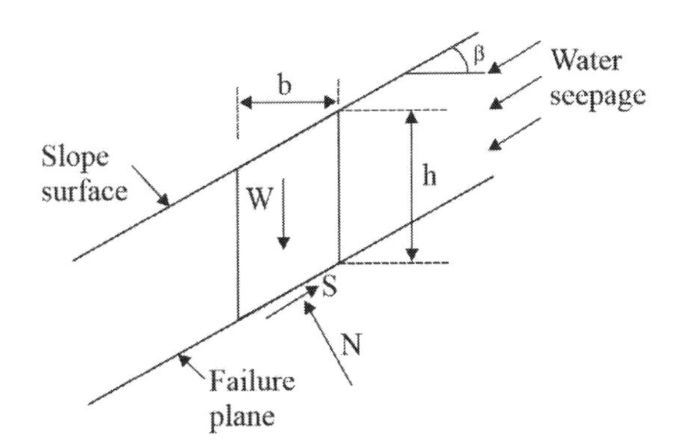

Figure 11.5 Forces acting on an infinite slope. *W* is the weight, *N* is the normal force, *S* is the available strength, *β* is the inclination of slope surface and *h* is the average thickness of the soil layer.

Solution:

For the 'dry' case, Equation 11.2 is used as follows:

$$FS = \frac{10}{20 \cdot 2 \cdot \sin 30^\circ \cdot \cos 30^\circ} + \frac{\tan 35^\circ}{\tan 30^\circ} = 0.58 + 1.21 = 1.79$$

Considering the seepage parallel to the slope (Equation 11.3), the factor of safety equals

$$FS = \frac{10}{20 \cdot 2 \cdot \sin 30^\circ \cdot \cos 30^\circ} + \left(\frac{20 - 9.81}{20}\right) \cdot \frac{\tan 35^\circ}{\tan 30^\circ} = 0.58 + 0.51 \cdot 1.21 = 1.2$$

11.3.2 Block failure

Hard rock is typically so strong that failure under gravity alone only occurs when discontinuities permit easy movement of discrete blocks. This may happen in bedded or foliated rock which are cut by joints. Block failure typically occurs when rock mass slides as one block. The slide forms under gravity alone when a rock block rests in an inclined weakness plane that 'daylights' into free space. This problem can be analyzed with a closed form solution (Figure 11.6) that depends on the slope geometry and the shear strength parameters of the rock material along the failure plane (Equation 11.4).

$$FS = \frac{2c \sin i}{\gamma H \sin \theta \sin(i - \theta)} + \frac{\tan \phi}{\tan \theta} \tag{11.4}$$

Problem 11.2: Rock slope in Figure 11.6 has a height of 12 m. A plane failure is expected along a large joint surface that has a friction angle of 33°. The plane has a dip of 42° while the slope angle is 64°. Considering the unit weight of 22 kN/m^3 and dry conditions, determine the safety factor.

Solution:

Equation 11.4 will be used to solve Problem 11.2.

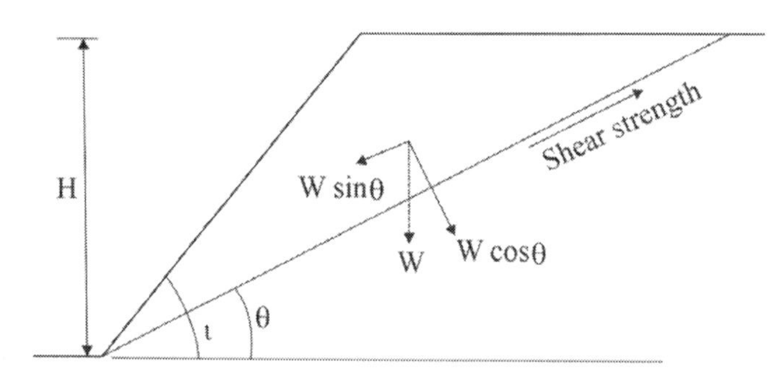

Figure 11.6 Block failure surface. W is the weight of the block, θ is the inclination of the failure plane, i is the slope inclination.

Question: *How can this equation be used if the cohesion is not given?*
Answer: There are times when we cannot obtain all data and we need to make some assumptions. In this case, we deal with a large joint surface between the potential landslide mass and bedrock. This means that there may not be cohesion and the shear strength is largely due to the surface friction. Thus, we will assume that $c = 0$, and the safety factor will be equal to

$$FS = \frac{\tan 33^\circ}{\tan 42^\circ} = \frac{0.649}{0.9} = 0.72$$

11.3.3 Wedge failure

Wedge failure occurs when rock mass slides as a wedge. It appears to be the most common type of failure in jointed rock mass when large joints intersect, forming a wedge. To estimate the possibility of wedge failure, we will consider rock slope geometry as schematically shown in Figure 11.7.

To compute the safety factor, the orientation of discontinuity/joint planes (Planes A and B in Figure 11.7a) will be assessed through the coefficient K (Equation 11.5):

$$K = \frac{\sin \theta_A + \sin \theta_B}{\sin(\theta_A + \theta_B)} \tag{11.5}$$

The safety factor will be calculated using Equation 11.6:

$$FS = K\left(\frac{\tan \phi}{\tan \beta}\right) \tag{11.6}$$

Problem 11.3: The rock slope shown in Figure 11.7 may be affected by wedge failure. There are two major joint planes that form the intersection angle with the slope's crest of 45° (θ_A) and 48° (θ_B), respectively. The line of intersection of these two planes has trend/plunge values of 047/53. The orientation of the slope face is described as 62/076. The friction angle along the joints is approximately 30° while the cohesion is 0. Determine the safety factor of this wedge.

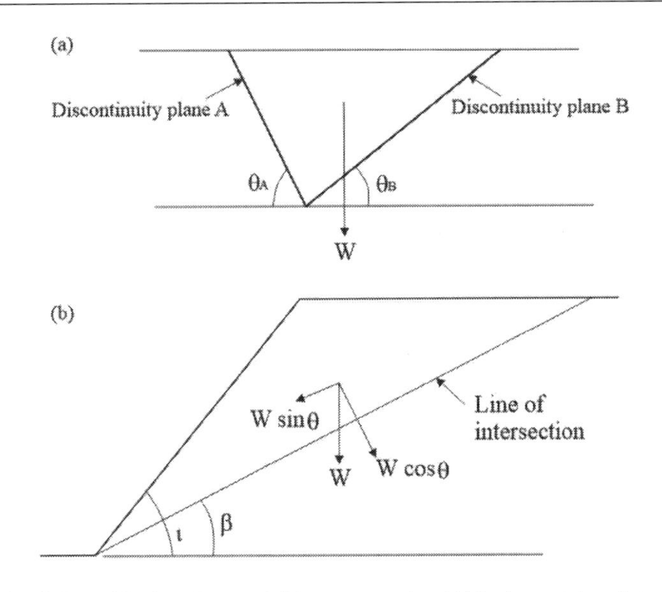

Figure 11.7 Wedge failure: (a) plan view, and (b) cross-section. W is the weight of the rock mass, β is the inclination of the line of intersection, i is the inclination of the slope, θ_A is the angle between the discontinuity plane A and the slope crest, and θ_B is the angle between the discontinuity plane B and the slope crest.

Solution:

From Equation 11.5, K can be found as follows:

$$K = \frac{\sin 45^\circ + \sin 48^\circ}{\sin(45^\circ + 48^\circ)} = \frac{0.707 + 0.743}{0.998} \approx 1.45$$

The safety factor is estimated using Equation 11.6:

$$FS = 1.45\frac{\tan 30^\circ}{\tan 53^\circ} = 1.45\frac{0.577}{1.32} \approx 0.63$$

11.4 Limit equilibrium method

The limit equilibrium method (LEM) is a complex analysis in which rock mass is divided into slices (Figure 11.8a) where the forces acting on each slice are considered (Figure 11.8b). There are a few assumptions made: (a) failure occurs along a distinct slip surface, (b) the sliding mass moves as an intact body, and (c) the Mohr-Coulomb failure criterion is satisfied along the slip surface.

Different assumptions regarding the direction and magnitude of the interslice forces result in different LEM methods including the Ordinary method of slice (Fellenius method), Bishop simplified method, Janbu method, Morgenstern-Price method and Spencer method.

Bishop's simplified method appears to be widely used in practice due to its simplicity and satisfactory results. The assumptions used in this method are:

1. The vertical interslice forces are equal and opposite
2. The interslice shear forces are zero.

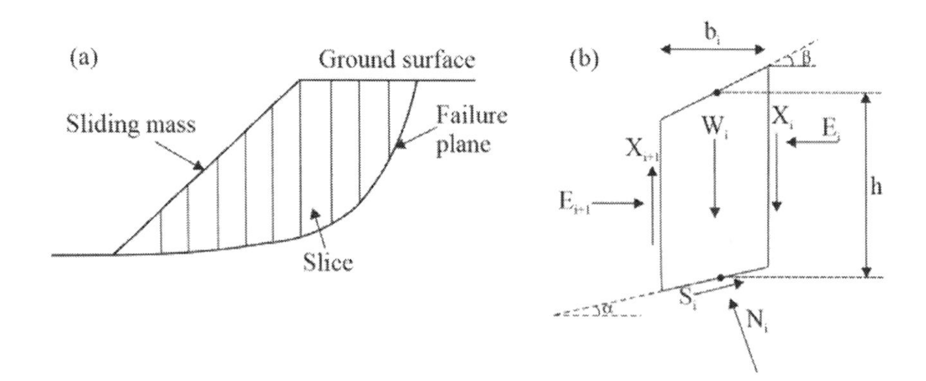

Figure 11.8 Division of potential sliding mass into slices (a), forces acting on a typical slice (b). *W* is the weight of slice, *N* is the normal force, *S* is the available strength, *E* and *X* are the interslice forces acting on the side of each slice, *b* is the width of slice, α is the inclination of slice base, β is the inclination of slice top and *h* is the average slice height.

Equations 11.7 and 11.8 give the mathematical solution of Bishop's simplified method where the pore water pressure is considered by using the effective shear strength parameters:

$$FS = \frac{Resisting\ moment}{Overturning\ moment} = \frac{\sum_{i=1}^{n}\left[c' \cdot b_i + \left(W_i - u_i \cdot b_i\right)\tan\phi'\right]\frac{1}{m_{\alpha(i)}}}{\sum_{i=1}^{n} W_i \cdot \sin\alpha_i} \tag{11.7}$$

where

$$m_{\alpha(i)} = \cos\alpha_i + \frac{\tan\phi' \cdot \sin\alpha_i}{FS} \tag{11.8}$$

As the term *FS* is present on both sides of Equation 11.7, a trial-and-error procedure is used to find the value of *FS*. A number of failure surfaces must be investigated so that the critical surface that provides the minimum factor of safety can be found. These iterative calculations are time-consuming and can easily take about an hour each. However, using a suitably programmed electronic spreadsheet or commercially available computer programs, the solutions can be determined rapidly.

11.5 Project work: slope stability analysis

The stability of the slope with no ground water during a dry season (case 1, the best-case scenario) will be analyzed first. The shear strength parameters of sandstone, mudstone and andesite rock mass are taken from Table 9.16. Commercially available software will be employed to determine the critical failure plane (i.e., the failure plane with the lowest safety factor) using Bishop's simplified method. Figure 11.9 presents the results of such an analysis, indicating that the safety factor is much greater than 1.

Question: *How reliable are these results?*

Answer: We need to remember that slope stability analysis involves theoretical assumptions and simplifications, which may be different from the existing stress and soil conditions

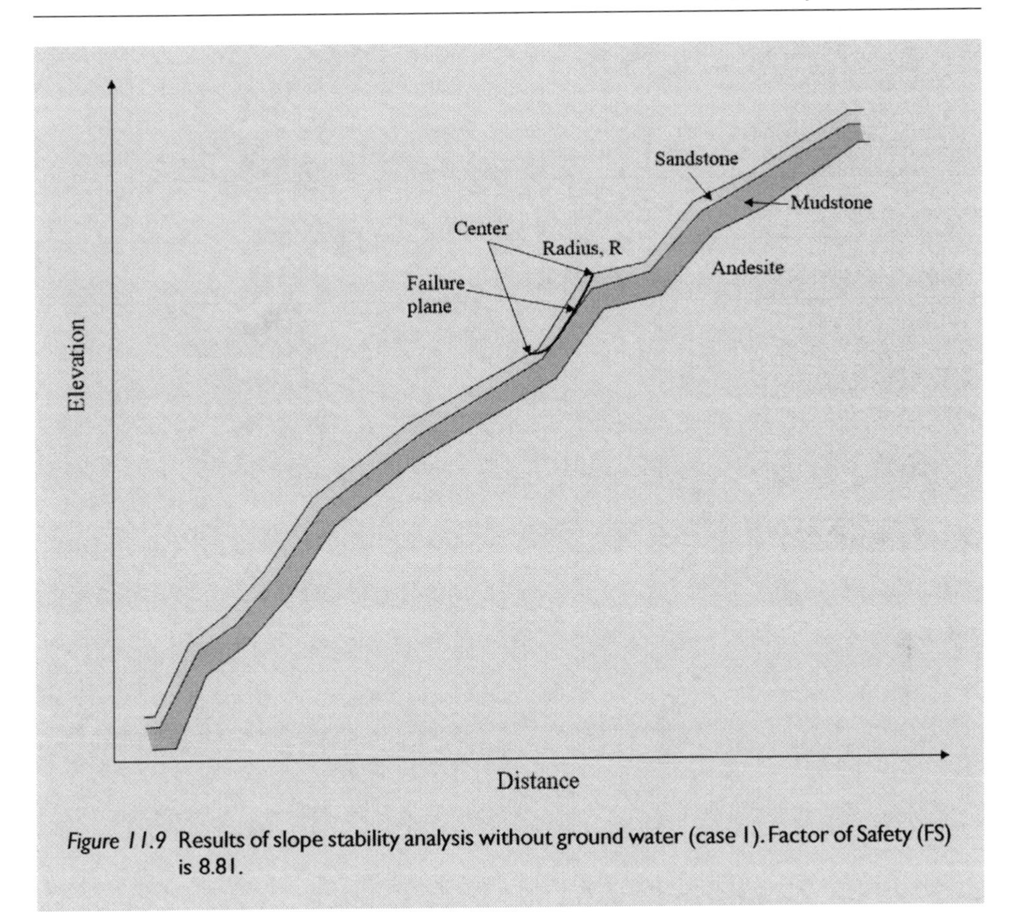

Figure 11.9 Results of slope stability analysis without ground water (case 1). Factor of Safety (FS) is 8.81.

in the field. For example, to create a model (Figure 11.9) we assume that the thickness of each layer does not vary much across the whole slope and the boundary between the geological units are parallel to the surface. This may not be correct as this assumption is based on the data from only four points (boreholes) across the 200 m long slope. For this reason, we should be cautious with the outcome from slope stability analysis.

The effect of ground water on the slope stability can be investigated as well. It will be assumed that the ground water level is about 1 m below the ground surface and because of water, the cohesion of the sandstone has dropped to 0 (case 2, the worst-case scenario). The results of the slope stability analysis for case 2 is given in Figure 11.10. It is evident from this figure that the location of the critical failure plane remains almost the same (compared to the one in Figure 11.9); however, the safety factor has significantly reduced to a value of 0.55, indicating slope instability.

Question: *The results show that the slope may become unstable. What shall we do?*
Answer: Indeed, during the worst-case scenario the slope may experience some stability issues; however, this scenario may not happen at any time soon or may not happen at all. If the stability of this slope is of high priority, some stabilization methods should be considered (this will be discussed in Section 11.7). Prior to that, it is wise to collect more information about the surface movement, which can be achieved through slope monitoring.

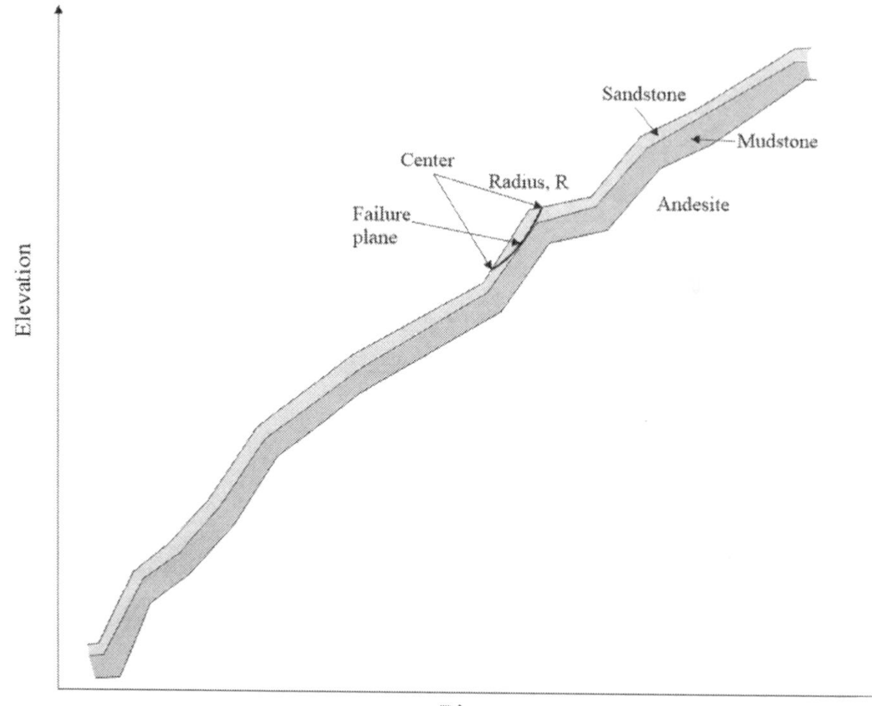

Figure 11.10 Slope stability analysis of the slope with ground water level of 1 m below the ground surface and the sandstone cohesion of 0 (case 2). Factor of Safety (FS) is 0.55.

11.6 Slope monitoring

Landslide monitoring is an important part of the landslide hazard management. The rate of landslide movement can be determined using either extensometers (surface movement, Figure 11.11) or inclinometers (movements at a certain depth, Figure 11.12). Inclinometers can also provide important information about the location of the failure plane/zone.

In some cases, when the rate of movement is rather low, monitoring can continue for several months or years (creep). If the rate of movement increases over time due to factors such as rainfall, weathering or human activities, it can lead to the occurrence of landslides. There are three types of landslide movements: regressive, progressive and transgressive (Figure 11.13). *Regressive*: when the deformation rate or slope velocity decreases with time from the original loading point, which can be either blasting, excavation or excessive pore pressures. This type of movement does not lead to collapse. *Progressive*: when the deformation rate or slope velocity increases with time, resulting in collapse. *Transgressive*: when the slope movements start as regressive but later become progressive (Figure 11.13).

11.7 Stabilization and protection techniques

Depending on the design concept, stabilization ('active' measures to prevent disaster) or protection/mitigation ('passive' measures that deal with the landslide consequences) can be considered. Table 11.3 summarizes common methods that deal with slope stability problems.

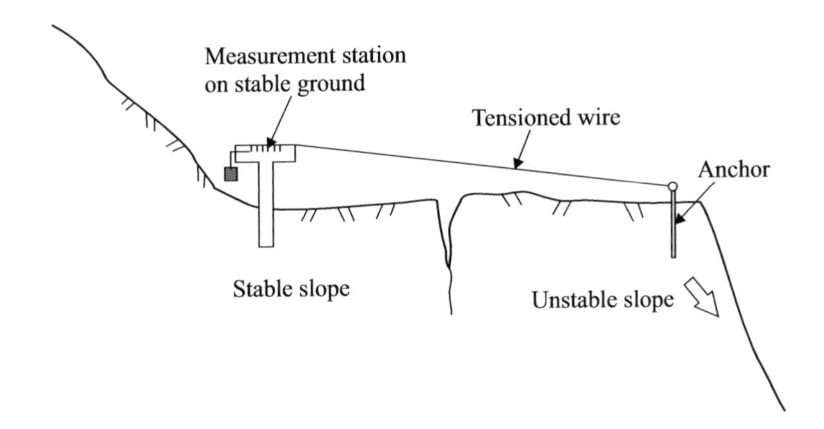

Figure 11.11 Extensometer setup to measure movements of unstable part of the slope

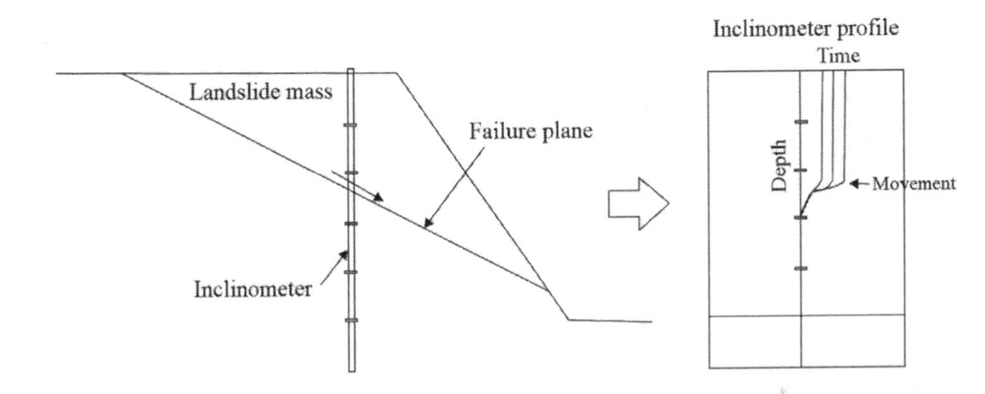

Figure 11.12 Inclinometer used to determine the depth of failure plane and monitor landslide movement over time

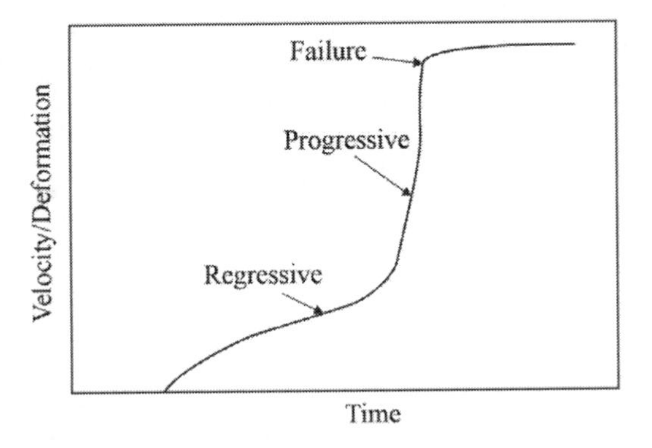

Figure 11.13 Example of transgressive deformations: rock mass movements start as regressive but later become progressive, resulting in failure

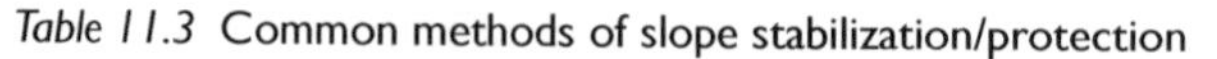

Table 11.3 Common methods of slope stabilization/protection

Type of measures	*Type of action*	*Description*
Stabilization – Rock removal	Scaling	Removal of loose rocks from the slope
	Resloping	Removal of weathered material is probably the most effective but expensive method. It is generally used for small to medium-sized slopes.
Stabilization – Reinforcement	Anchors and piles	*Anchors*. This method utilizes the tensile force of anchor bodies, which are placed throughout the sliding mass and fixed in the bed rocks (Figure 11.14). *Piles* are used to 'tie' the sliding mass with the bed rocks below to restrain the movement. Steel piles filled with concrete are typically used for this purpose.
	Shotcrete	Shotcrete protects the slope from being saturated during rainfall and it also adds some stability to the slope face.
Protection	Drainage	Surface drainage ditches are designed to collect and remove rain water. It is one of the most effective and relatively inexpensive methods. However, it needs to be regularly maintained, especially in a forested area (Figure 11.15).

Figure 11.14 Rock slope stabilized with a set of anchors and shotcrete

Question: *What are the major principles of slope stabilization/protection methods?*
Answer: The slope stability analysis we conducted in Section 11.5 shows the adverse effect of water on rock strength. Also, the effect of rock strength on slope stability can be seen when the cohesion of rock material significantly decreases, making the slope less stable. Thus, the main concepts of stabilization techniques are to prevent water from going in the

Figure 11.15 Example of surface drainage (clogged with soil and leaves) that has not been maintained for a long time

slope (surface drainage or shotcrete) and increasing the rock strength by means of anchors. Another way to increase the safety factor is to decrease the shear stress acting on the potential landslide mass, which can be achieved by reshaping the slope surface.

11.8 Problems for practice

Problem 11.4: The slope shown in Figure 11.6 is under critical conditions. The height of the slope is 12 m. The failure plane is formed in weathered sandstone. The failure plane is inclined at an angle of 45°. The slope face has an orientation of 55/065. The rock density is 2.3 g/cm^3. If the cohesion of the sandstone along the failure plane is 10 kPa, what will be the friction angle?

Solution:

We know that

$$c = 10 \text{ kPa}, \theta = 45°, i = 55°, H = 12 \text{ m}.$$

The unit weight of rock is

$$\gamma = 2.3 \cdot 9.81 \approx 22.56 \; kN \; m^{-3}$$

For the critical conditions, the factor of safety is $FS = 1$.
Then, we will have

$$1 = \frac{2 \cdot 10 \cdot \sin 55^{\circ}}{22.56 \cdot 12 \cdot \sin 45^{\circ} \cdot \sin(55-45)} + \frac{\tan\phi}{\tan 45^{\circ}}$$

From this relationship, we will obtain the friction angle as

$$\phi \approx 27^{\circ}$$

Problem 11.5: For the rock slope in Figure 11.7, there are two major joint planes that form the intersection angle with the slope's crest of 43° and 49°, respectively. The line of intersection of these two planes has trend/plunge values of 046/50. The orientation of the slope face is 59/063. What is the minimum value of the friction angle that will keep this slope stable?

Solution:

From Figure 11.7, we will have

$$\theta_A = 43°, \theta_B = 49°, \beta = 50°, \alpha = 59°$$

The coefficient K can be obtained using Equation 11.5 as follows:

$$K = \frac{\sin 43^{\circ} + \sin 49^{\circ}}{\sin(43+49)} = \frac{0.681 + 0.754}{0.999} \approx 1.44$$

The minimum value of friction angle that will keep this slope stable is associated with the critical conditions (i.e., $FS = 1$). Then, we will have (Equation 11.6)

$$1 = 1.44 \cdot \frac{\tan\phi}{\tan 50^{\circ}}$$

This relationship will result in

$$\phi \approx 39.6^{\circ}$$

11.9 Review questions

1. There are three major categories of landslide causes such as geological, human, and

   a) topographical   b) geometrical   c) geographical   d) morphological

2. Which of the following landslide causes is not considered to be a geological cause?

   a) weak materials   b) weathered materials
   c) subterranean erosion   d) adversely oriented discontinuities

3. Which of the following is not a cause of slope failure?

   a) rainfall   b) wind   c) erosion   d) human activities

4. Lateral spreads typically occur in large rock masses which are inclined in the range of

   a) <10°   b) 30°–40°   c) 60°–70°   d) > 70°

5. Which assumption is NOT used in limit equilibrium analysis?

   a) Failure occurs along a distinct slip surface
   b) For the sliding body, all equations of static equilibrium are satisfied
   c) Mohr-Coulomb failure criterion is satisfied along the slip surface
   d) The sliding mass moves as separate slices

6. The factor of safety of a rock slope for block failure does not depend on the

   a) width of the slope   b) height of the slope
   c) inclination of the slope   d) cohesion of the rock mass

7. The factor of safety (FS) is defined as:

   a) FS = Available strength/Required strength
   b) FS = Required strength/Available strength

8. When FS = 1.5, it means

   a) 100% strength mobilization   b) 67% strength mobilization
   c) 50% strength mobilization   d) 33% strength mobilization

9. Bishop's slope stability analysis is based on

   a) Limit analysis   b) Limit equilibrium method
   c) Finite element analysis   d) Boundary element analysis

Answers: 1. d   2. c   3. b   4. a   5. d   6. a   7. a   8. b   9. b

Chapter 12

Rocks and tunnels

Project relevance: This chapter will discuss engineering problems related to rock mass during underground constructions. Rockbursting or squeezing can occur in hard or soft rocks under high stresses, compromising the stability of underground structures. The main factors causing instability issues in different geological units will be discussed. In addition, engineering problems that may occur at the project site during underground excavation will be analyzed and assessed.

12.1 Behavior of rock mass in tunnels

The principal factors that are generally considered in tunnel design are the in situ stresses and the mechanical properties of the rock mass. At greater depths, the rock mass can be overstressed, resulting in either brittle fracture of the intact rock or shear failure along preexisting discontinuities such as joints or shear zones. These two types of failure can coexist and the extent to which the failures propagate depends upon the characteristics of the rock mass, the magnitude and directions of the in situ stresses, the shape of the tunnel, and the intensity and orientation of the discontinuities (Hoek and Marinos, 2009).

Depending on the stress level and strength of rock, there are a few major engineering problems that can occur during tunnel excavation or mining. Table 12.1 provides a brief description of each type.

12.2 Brittle failure of massive rock mass

Extremely hard and massive rock mass is preferable environment for tunnel construction; however, some engineering problems may occur at greater depths where the rock mass is under very high stresses. *Rockbursts* and *spalling* (Figure 12.1) typically occur in high strength rock mass (> 100 MPa) with few or no discontinuities at depths of more than 1000 m. Rock popping commonly happens from the side or the roof of the tunnel where the rock is in a state of intense elastic deformation. Spalling refers to rock formations from which slabs of rocks are suddenly detached after the rock has been exposed in a tunnel.

The aforementioned engineering problems are encountered only in hard and brittle rock where the maximum stress on the tunnel boundary exceeds about 40% of the uniaxial compressive strength of the rock. It is assumed that at the wall of the tunnel, the stress in the rock

Table 12.1 Common problems in rock mass during tunnel excavation

Problem	*Description*	*Rock mass structure (according to GSI)*	*Comments*
Rockburst	It is a sudden, violent rupture or collapse of highly stressed rock	Intact or massive rock mass	Instability of hard massive rock mass under very high stresses
Spalling	When rock mass gradually breaks into pieces, flakes or fragments under high stresses		
Wedge/block failure	It is block failure, where pre-existing blocks in the roof and side walls become free to move because of the excavation	Very blocky rock mass	Mainly gravity-driven, structurally dependent instability under relatively low stresses
Raveling ground	This phenomenon refers to the falling out of individual rock blocks due to overstressing	Disintegrated rock mass with very good to good surface quality	Stress dependent instabilities under relatively high stresses with a strong influence of discontinuities
Squeezing	Plastic flow of incompetent rock into the tunnel under the influence of high pressure	Disintegrated rock mass with poor to very poor surface quality	Deformation problems caused by relatively low strength of the rock mass

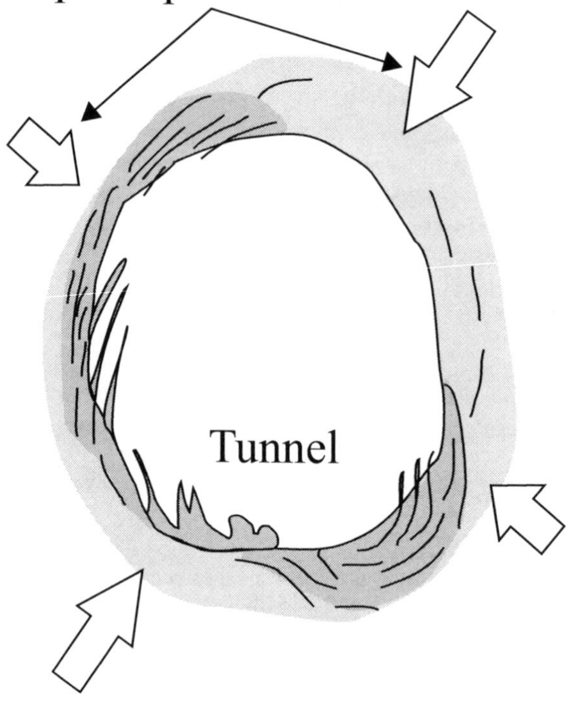

Figure 12.1 Brittle failure of strong massive rock under high in situ stress levels

is similar to that in an unconfined rock specimen subject to axial compression in a testing machine.

Question: *What can happen when a tunnel is excavated in stratified (bedded) rocks?*
Answer: Stratified rock breaks readily along the bedding planes because the bedding planes constitute a source of mechanical weakness. In foliated rocks like schist, the cleavage planes can have a similar effect.

Tensile cracks typically initiate at 40%–50% of the uniaxial compressive strength of most massive rock. The occurrence of cracks can be estimated using the relationship between the brittle crack initiation stress and uniaxial compressive strength of massive rock (Hoek and Marinos, 2009).

$$\sigma_{init} = -2.7 + 0.42 \cdot \sigma_c \tag{12.1}$$

where σ_{init} is the crack initiation stress and σ_c is the uniaxial compressive stress of rock.

Question: *How do engineers deal with these problems?*
Answer: It is necessary to provide tunnel support during excavation, which may include rock bolts and steel sets. However, for some massive rock masses, it becomes extremely difficult to support the tunnel when the ratio of the maximum boundary stress to UCS increases, as shown in Figure 12.2.

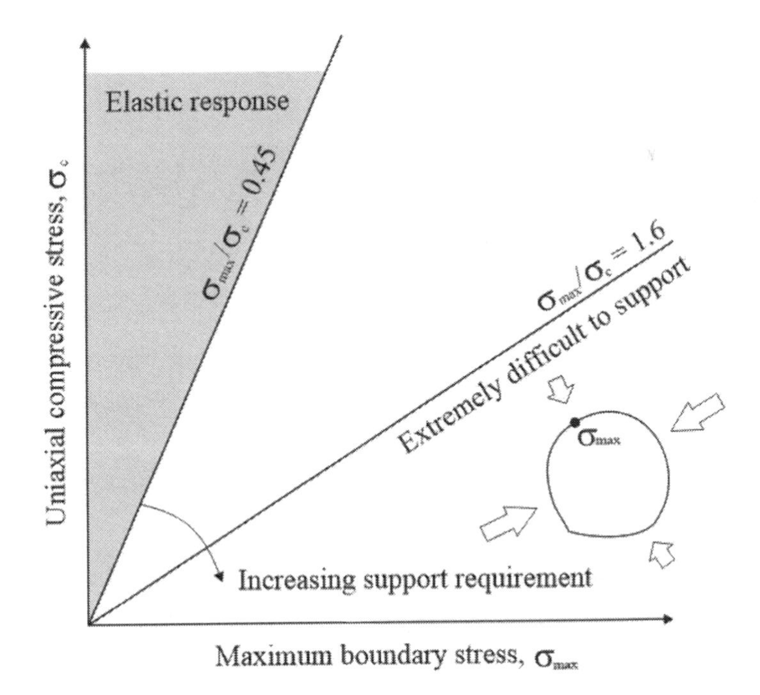

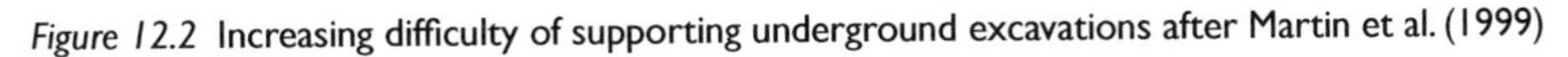
Figure 12.2 Increasing difficulty of supporting underground excavations after Martin et al. (1999)

12.3 Gravitational failure

When tunnels are excavated in jointed rock mass at relatively shallow depths, the most common type of failure is wedge and/or slide of rock blocks. The wedges are formed due to the intersection of structural features such as bedding planes and joints. When a free face is created during tunnel excavation, the support from the neighboring rock is removed and the wedge can easily fall or slide into the opening. Gravitational failures schematically shown in Figure 12.3 include rock block sliding from the wall or free fall from the roof of tunnel.

12.4 Problems with disintegrated rock mass

The problem of raveling ground can occur in disintegrated rock mass under relatively high stress conditions. It occurs when a tunnel is excavated through large shear zones or fault where the strength of rock mass is greatly affected by numerous discontinuities.

Another problem related to disintegrated rock mass is known as *squeezing*. It can occur during tunnel excavation in relatively weak rock mass where the low-strength rock forms the plastic zone around the tunnel (Figure 12.4), resulting in radial convergence (deformation).

Question: *Both raveling ground and squeezing occur in disintegrated rock mass. Do they both occur in the same type of rock?*
Answer: No. Raveling ground occurs in high strength but disintegrated rock mass affected by structural discontinuities such as faults, while squeezing occurs in soft (low strength) rock that often contains considerable amount of clay (for example, shales).

Site measurements and monitoring of some tunnel sites in Taiwan (Chern et al., 1998) with squeezing issues indicate that the strain level of more than 1% is commonly associated with the onset of tunnel instability.

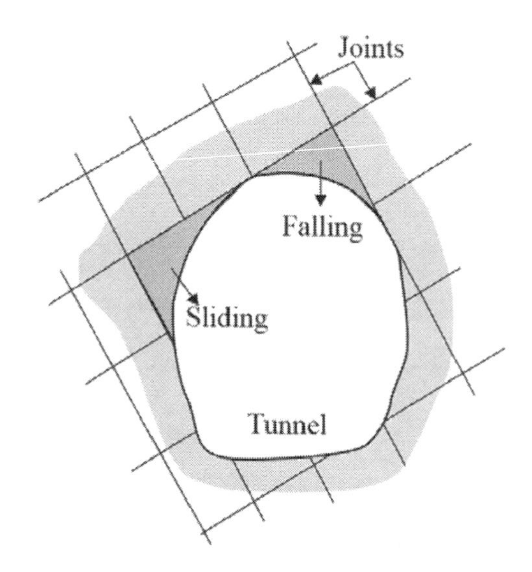

Figure 12.3 Gravitational failure of blocks or wedges defined by intersecting of structural features. Such failures typically occur in very blocky rock mass.

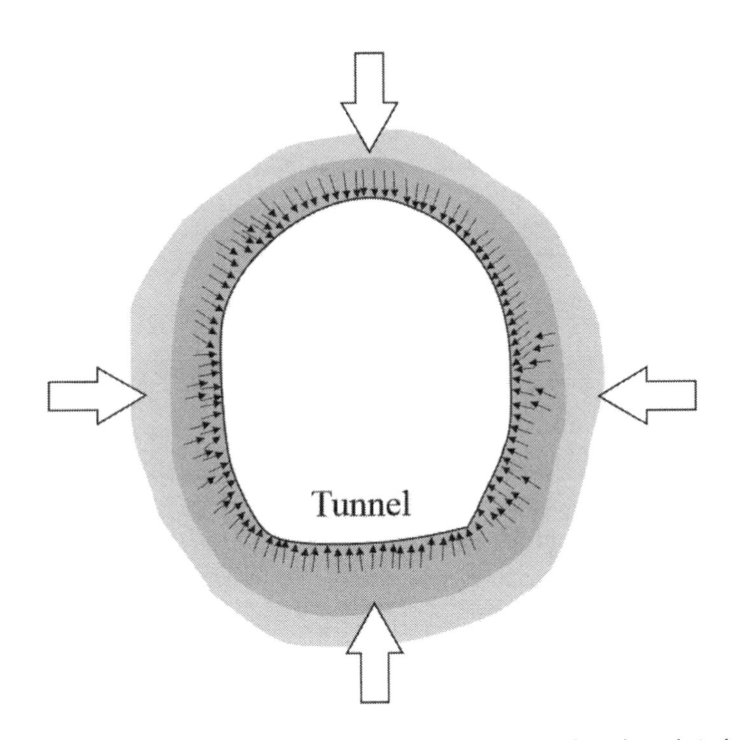

Figure 12.4 Formation of the 'plastic' zone by shear failure of weak rock under relatively high stress compared to the strength of the rock mass

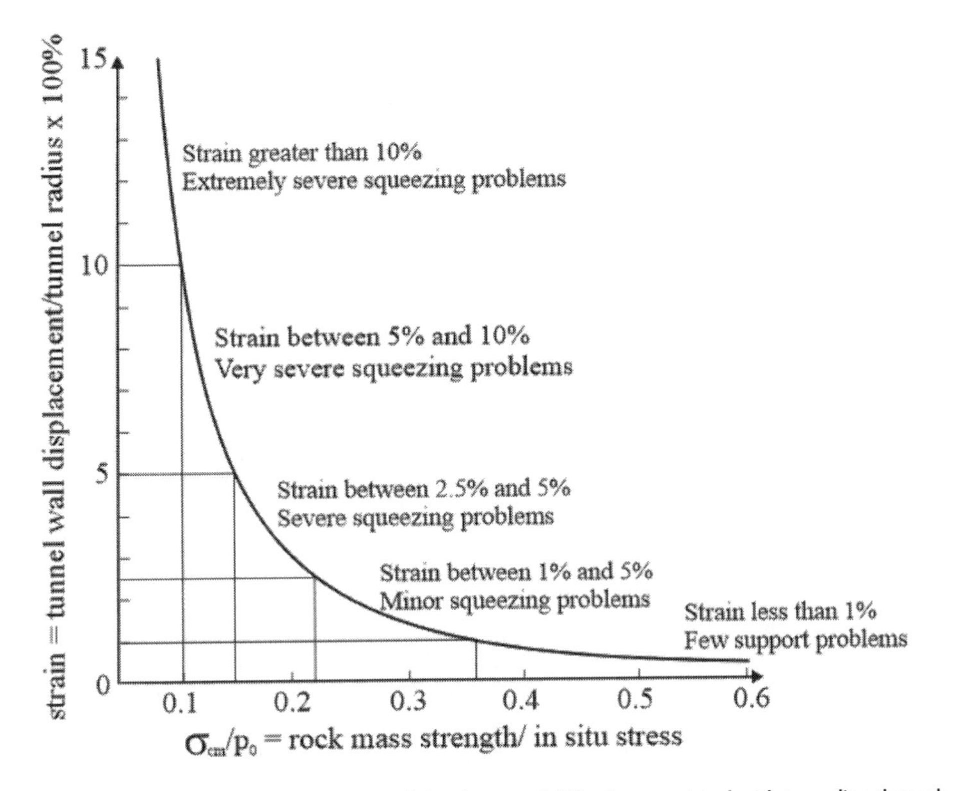

Figure 12.5 Relationship between strain and the degree of difficulty associated with tunneling through squeezing rock (after Hoek and Guevara, 2009)

Hoek (1998) suggested that the ratio of the uniaxial compressive strength (σ_{cm}) of the rock mass to the in situ stress p_0 can be used as an indicator of potential tunnel squeezing problems. Figure 12.5 gives estimates of the percentage strain which is defined as the ratio of tunnel closure to tunnel diameter in respect to the strength/stress ratio. It is also suggested that the strain (ε) can be approximately estimated using Equation 12.2.

$$\varepsilon = 0.2\left(\frac{\sigma_{cm}}{p_0}\right)^{-2} \tag{12.2}$$

Question: *What is the practical application of the plot in Figure 12.5?*
Answer: This plot allows to estimate the expected strain of the tunnel wall, which assists with selecting the most appropriate method of stabilization. Hoek (2007) suggests the following measures, which are briefly summarized in Table 12.2.

12.5 Project work: tunnel-related problems

Some potential problems (i.e., crack initiation and squeezing) that may occur in the andesite and mudstone during tunnel construction will be estimated.

Tunnel in the andesite at a depth of 19 m

The crack initiation stress for this andesite equals

$$\sigma_{init} = -2.7 + 0.42 \cdot \sigma_c = -2.7 + 0.42 \cdot 60 = 22.5\,MPa$$

The average unit weight of andesite is 26 kN/m^3 (Table 7.2). The stress at a depth of 19 m will be

$$\sigma = \gamma \cdot z = 26 \cdot 19 = 494\ kPa = 0.49\ MPa$$

Table 12.2 Support types in tunnels with poor quality of rock

Strain (%)	*Support type*
Less than 1	Very simple tunneling conditions with rockbolts and shotcrete typically used for support
1 to 2.5	Minor squeezing problems which are generally dealt with rockbolts and shotcrete
2.5–5	Severe squeezing problems that require rapid support installation and control of construction quality. Heavy steel sets embedded in shotcrete are generally required.
5–10	Very severe squeezing and face stability problems. Forepolling and face reinforcement with steel sets embedded in shotcrete are usually necessary.
More than 10	Extreme severe squeezing problems. Forepolling and face reinforcement are usually applied and yielding support may be required in extreme cases.

The stress at a depth of 19 m is not sufficient ($\sigma << \sigma_{init}$) to cause rockburst or spalling in the andesite.

Tunnel in the mudstone at a depth of 7 m

The average unit weight of mudstone is 20.9 kN/m^3 (Table 7.2). The stress at a depth of 7 m equals

$$\sigma = 20.9 \cdot 7 \approx 146.3\ kPa = 0.146\ MPa$$

The rock mass strength (σ_{cm}) is 6.73 MPa (Table 9.16), and the in situ stress (p_0) is 0.146 MPa. The ratio between these two parameters gives

$$\frac{\sigma_{cm}}{p_0} = \frac{6.73}{0.146} \approx 46.1$$

This value implies that no squeezing problems will occur during tunnel construction (Figure 12.5). Indeed, the stress level is rather low compared to the strength of rock mass, which will help avoid squeezing and rockburst problems.

12.6 Problems for practice

Problem 12.1: A tunnel will be constructed in hard granite (σ_c = 100 MPa, density is 2.9 g/cm^3). At what depth could the problem of rockburst or spalling occur?

Solution:

The crack initiation stress can be estimated using Equation 12.1 and σ_c = 100 MPa:

$$\sigma_{init} = -2.7 + 0.42 \cdot 100 = 39.3\ MPa$$

The depth (z) at which this stress (39.3 MPa = 39.3·10^3 kPa) exists can be found as follows:

$$\sigma = \sigma_{init} = \gamma \cdot z$$

$$z = \frac{\sigma_{init}}{\gamma} = \frac{39.3 \cdot 10^3}{2.9 \cdot 9.81} \approx 1.38 \cdot 10^3\ m = 1380\ m$$

Problem 12.2: A tunnel will be constructed in gneiss at a depth of 1500 m. The rock's density is 3.1 g/cm^3 and unconfined compressive strength is 40 MPa. What problems may occur during construction?

Solution:

The unit weight of gneiss equals

$$\gamma = 3.1 \cdot 9.81 \approx 30.4\ kN \cdot m^{-3}$$

The maximum stress at a depth of the tunnel can be estimated as

$$\sigma_{max} = \gamma \cdot z = 0.03 \cdot 1500 \approx 45\ MPa$$

The ratio between the stress and UCS (σ_c) of rock will be

$$\frac{\sigma_{max}}{\sigma_c} = \frac{45}{40} \approx 1.13$$

According to Figure 12.2, the construction needs support.

Problem 12.3: Tunnels in poor quality rocks. A tunnel will be excavated at a depth of 44 m in flysch with the estimated rock mass strength of 0.17 MPa. What problems may occur during the construction?

Solution:

To estimate the stress at a depth of 44 m, Equation 8.2 will be used.

$$\sigma = 0.027 \cdot 44 \approx 1.19\ MPa$$

The rock mass strength (σ_{cm}) is 0.17 MPa, while the in situ stress (p_0) is equal to 1.19 MPa. The ratio between these two parameters is

$$\frac{\sigma_{cm}}{p_0} = \frac{0.17}{1.19} \approx 0.14$$

According to Figure 12.5, this stress ratio correlates with the percentage strain that is equal or greater than 5%. This implies severe squeezing problems during tunneling.

12.7 Review quiz

1. The principal factors that need to be considered for a tunnel design are the mechanical properties of the rock mass and

   a) in situ stress b) joints of rocks
   c) rock porosity d) the tunnel length

2. For a tunnel constructed in rock mass with massive structure (according to GSI) at a depth of 1500 m, what problem can typically occur?

   a) rockburst b) squeezing c) swelling d) shear failure

3. In a tunnel constructed in granites at a depth of 1500 m, the common type of rock failure would be

   a) brittle b) ductile c) brittle + ductile d) plastic failure

4. Spalling typically occurs in rock mass described by GSI as

   a) massive b) blocky c) very blocky d) disturbed

5. What problem would typically occur in *disintegrated* rock mass (according to GSI) during tunnel excavation under relatively high stress conditions?

   a) raveling ground b) spalling c) squeezing d) wedge failure

6. In tunnels, structural failures such as fall or slide of wedges typically occur in rock mass described by GSI as

   a) massive b) blocky c) very blocky d) disturbed

7. Rockbursts typically occur in rock masses with the strength (UCS) of

   a) <20 MPa b) 20–50 MPa c) 50–80 MPa d) >100 MPa

Answers: 1. a 2. a 3. a 4. a 5. a 6. c 7. d

References

American Society for Testing and Materials. (2001) Standard test method for determination of rock hardness by rebound hammer method. ASTM Stand., 04.09 (2001) (D 5873-00), ASTM, Philadelphia.

American Society for Testing and Materials. (2008) Standard test method for determination of the point load strength index of rock and application to rock strength classifications (D5731). *Annual Book of Standards* Volume 04.08, ASTM, Philadelphia.

Atkinson, L. C. (2000). The role and mitigation of groundwater in slope stability. *Slope Stability in Surface Mining*, 427–434.

Barton, N. (1973) Review of a new shear-strength criterion for rock joints. *Engineering Geology*, 7(4), 287–332.

Barton, N. (1978) Suggested methods for the quantitative description of discontinuities in rock masses. *ISRM, International Journal of Rock Mechanics and Mining Sciences & Geomechanics Abstracts*, 15(6), 319–368.

Barton, N. (1995) The influence of joint properties in modelling jointed rock masses. In 8th ISRM Congress. International Society for Rock Mechanics.

Barton, N. (2007) Rock mass characterization for excavations in mining and civil engineering. In *Proceeding of the International Workshop on Rock Mass Classification in Underground Mining*. IC Volume 9498, pp. 3–13.

Barton, N. & Choubey, V. (1977). The shear strength of rock joints in theory and practice. *Rock Mechanics*, 10(1–2), 1–54.

Barton, N., Lien, R. & Lunde, J. (1974) Engineering classification of rock masses for the design of tunnel support. *Rock Mechanics*, 6(4), 189–236.

Bieniawski, Z.T. (1968) The effect of specimen size on compressive strength of coal. *International Journal of Rock Mechanics and Mining Sciences & Geomechanics Abstracts*, 5(4), 325–335. Pergamon.

Bieniawski, Z.T. (1973) Engineering classification of jointed rock masses. *Civil Engineer in South Africa*, 15(12).

Bieniawski, Z.T. (1989) *Engineering Rock Mass Classifications: A Complete Manual for Engineers and Geologists in Mining, Civil, and Petroleum Engineering*. John Wiley & Sons, USA.

Broch, E. & Franklin, J.A. (1972) The point-load strength test. *International Journal of Rock Mechanics and Mining Sciences & Geomechanics Abstracts*, 9(6), 669–676. Pergamon.

Chern, J.C., Yu, C.W. & Shiao, F.Y. (1998) Tunnelling in squeezing ground and support estimation. *Proceedings of Regional Symposium Sedimentary Rock Engineering*, Taipei. pp. 192–202.

Cogan, J., Gratchev, I. & Wang, G. (2018) *Rainfall-induced Shallow Landslides Caused by Ex-Tropical Cyclone Debbie, 31st March 2017, Landslides*. pp. 1–7.

Cui, C., Gratchev, I., Chung, M. & Kim, D.H. (2019) Changes in joint surface roughness of two natural rocks during shearing. *International Journal of Geomate*, 17(63), 181–186.

Deere, D.U. (1964) Technical description of rock cores for engineering purpose. *Rock Mechanics and Engineering Geology*, 1(1), 17–22.

Descoeudres, F., Montani Stoffel, S., Böll, A., Gerber, W. & Labiouse, V. (1999) Rockfalls. In: Coping study on disaster resilient infrastructure. IDNDR, Zurich, pp. 37–47.

Franklin, J.A. & Chandra, R. (1972) The slake-durability test. *International Journal of Rock Mechanics and Mining Science*, 9(3), 325–341.

Goodman, R.E. (1989) *Introduction to Rock Mechanics*, Volume 2. Wiley, New York.

Goodman, R.E. (1993) *Engineering Geology*. John Wiley & Sons, USA.

Gratchev, I.B. & Towhata, I. (2011) Analysis of the mechanisms of slope failures triggered by the 2007 Chuetsu Oki earthquake. *Geotechnical and Geological Engineering*, 29(5), 695.

Gratchev, I.B., Sassa, K., Osipov, V.I. & Sokolov, V.N. (2006) The liquefaction of clayey soils under cyclic loading. *Engineering Geology*, 86(1), 70–84.

Gratchev, I. & Jeng, D.S. (2018) Introducing a project-based assignment in a traditionally taught engineering course. *European Journal of Engineering Education*, 43(5), 788–799.

Gratchev, I. & Kim, D.H. (2016) On the reliability of the strength retention ratio for estimating the strength of weathered rocks. *Engineering Geology*, 201, 1–5.

Gratchev, I. & Saeidi, S. (2019) The effect of surface irregularities on a falling rock motion. *Geomechanics and Geoengineering*, 14(1), 52–58.

Gratchev, I. & Towhata, I. (2010) Geotechnical characteristics of volcanic soil from seismically induced Aratozawa landslide, Japan. *Landslides*, 7(4), 503–510.

Gratchev, I., Irsyam, M., Towhata, I., Muin, B. & Nawir, H. (2011) Geotechnical aspects of the Sumatra earthquake of September 30, 2009, Indonesia. *Soils and Foundations*, 51(2), 333–341.

Gratchev, I., Kim, D.H. & Chung, M. (2015) Study of the interface friction between mesh and rock surface in drapery systems for rock fall hazard control. *Engineering Geology*, 199, 12–18.

Gratchev, I., Kim, D.H. & Yeung, C.K. (2016) Strength of rock-like specimens with pre-existing cracks of different length and width. *Rock Mechanics and Rock Engineering*, 49(11), 4491–4496.

Gratchev, I., Oh, E. & Jeng, D.S. (2018) *Soil Mechanics Through Project-Based Learning*. CRC Press, London.

Gratchev, I., Pathiranagei, S.V. & Kim, D.H. (2019) Strength properties of fresh and weathered rocks subjected to wetting – drying cycles. *Geomechanics and Geophysics for Geo-Energy and Geo-Resources*, 1–11.

Gratchev, I., Shokouhi, A., Kim, D.H., Stead, D. & Wolter, A. (2013) Assessment of rock slope stability using remote sensing technique in the Gold Coast area, Australia. In *18th Southeast Asian Geotechnical & Inaugural AGSSEA Conference*.

Griffith, J.H. (1937) Physical properties of typical American rocks. – Iowa Eng. Exper. *Station Bull*, 131, (35), 42.

Handin, J. (1966) Strength and ductility. S.P. Clark Jr. (Ed.), *Handbook of Physical Constants*. Geological Society of America, 97, pp. 238–289

Hoek, E. (1994). Strength of rock and rock masses. *ISRM News Journal*, 2(2), 4–16.

Hoek, E. (1998) Keynote address: tunnel support in weak rock. In: *Proceedings of the Symposium of Sedimentary Rock Engineering*, Taipei, Taiwan, 1998. pp. 1–12.

Hoek, E. (2007) *Practical Rock Engineering*. Online. ed. Rocscience. https://www.rocscience.com/assets/resources/learning/hoek/Practical-Rock-Engineering-Full-Text.pdf

Hoek, E. & Brown, E.T. (1980) Empirical strength criterion for rock masses. *Journal of Geotechnical and Geoenvironmental Engineering*, 106(ASCE 15715).

Hoek, E. & Brown, E.T. (1997) Practical estimates of rock mass strength. *International Journal of Rock Mechanics and Mining Sciences*, 34(8), 1165–1186.

Hoek, E. & Guevara, R. (2009) Overcoming squeezing in the Yacambú-Quibor tunnel, Venezuela. *Rock Mechanics and Rock Engineering*, 42(2), 389–418.

Hoek, E. & Marinos, P.G. (2000) Predicting tunnel squeezing problems in weak heterogeneous rock masses. *Tunnels and Tunnelling International*, 32(11), 45–51.

Hoek, E. & Marinos, P.G. (2009) Tunnelling in overstressed rock. In *Proceedings of the Regional Symposium of the International Society for Rock Mechanics, EUROCK*. pp. 49–60.

Hudson, J.A. & Priest, S.D. (1979) Discontinuities and rock mass geometry. *International Journal of Rock Mechanics and Mining Sciences & Geomechanics Abstracts*, 16, (6), 339–362. Pergamon.

ISRM (1981) Basic geotechnical description of rock masses. *International Journal of Rock Mechanics and Mining Sciences & Geomechanics Abstracts*, 18, 85–110.

Jumikis, A.R. (1983) *Rock Mechanics*. Gulf Publishing Company, Houston, p. 613.

Kim, D.H., Gratchev, I. & Balasubramaniam, A. (2013) Determination of joint roughness coefficient (JRC) for slope stability analysis: a case study from the Gold Coast area, Australia. *Landslides*, 10(5), 657–664.

Kim, D.H., Gratchev, I. & Balasubramaniam, A. (2015a). A photogrammetric approach for stability analysis of weathered rock slopes. *Geotechnical and Geological Engineering*, 33(3), 443–454.

Kim, D.H., Gratchev, I. & Balasubramaniam, A. (2015b) Back analysis of a natural jointed rock slope based on the photogrammetry method. *Landslides*, 12(1), 147–154.

Kim, D.H., Gratchev, I., Berends, J. & Balasubramaniam, A. (2015) Calibration of restitution coefficients using rockfall simulations based on 3D photogrammetry model: A case study. *Natural Hazards*, 78(3), 1931–1946.

Kim, D.H., Gratchev, I., Hein, M. & Balasubramaniam, A. (2016) The application of normal stress reduction function in tilt tests for different block shapes. *Rock Mechanics and Rock Engineering*, 49(8), 3041–3054.

Kim, D.H., Poropat, G., Gratchev, I. & Balasubramaniam, A. (2016) Assessment of the accuracy of close distance photogrammetric JRC data. *Rock Mechanics and Rock Engineering*, 49(11), 4285–4301.

Laubscher, D.M. & Page, C.H. (1990) The design of rock support in high stress or weak rock environments. *Proceeding of the 92nd Canadian Institute of Mining Metallurgy*. AGM.

Look, B.G. & Griffiths, S.G. (2001) An engineering assessment of the strength and deformation properties of Brisbane rocks. *Australian Geomechanics*, 36(3), 17–30.

Martin, C.D., Kaiser, P.K. & McCreath, D.R. (1999) Hoek-Brown parameters for predicting the depth of brittle failure around tunnels. *Canadian Geotechnical Journal*, 36(1), 136–151.

Meehan, R.L., Dukes, M.T. & Shires, P.O. (1975) A case history of expansive claystone damage. *Journal of Geotechnical and Geoenvironmental Engineering*, 101(ASCE# 11590 Proceeding).

Palmström, A. (1982) The volumetric joint count-a useful and simple measure of the degree of rock jointing. *Proceeding of the 4th Congress of International Association of Engineering Geology, 5*, 221–228.

Palmström, A. (2005) Measurements of and correlations between block size and rock quality designation (RQD). *Tunnelling and Underground Space Technology*, 20(4), 362–377.

Peck, R.B., Hanson, W.E. & Thornburn, T.H. (1974) *Foundation Engineering*, Volume 10. Wiley, New York.

Pierson, L.A. (1992) Rockfall hazard rating system. In: Rockfall prediction and control and landslides case histories. Transportation Research Record No. 1343, Washington, DC, pp. 6–13.

Priddle, J., Lacey, D., Look, B. & Gallage, C. (2013) Residual soil properties of South East Queensland. *Australian Geomechanics Journal*, 48(1), 67–76.

Proctor, R.V. & White, T.L. (1946) *Rock Tunneling With Steel Supports*. Commercial Shearing & Stamping Company.

Rapp, A. (1960) Recent development of mountain slopes in Kärkevagge and surroundings, Northern Scandinavia. *Geografiska Annaler*, 42(2–3), 65–200.

Ritchie, A.M. (1963) Evaluation of rockfall and its control. *Highway Research Record*, 17, 13–28.

Schmidt, E. (1951) A non-destructive concrete tester. *Concrete*, 59, 34–35.

Sherard, J.L., Cluff, L.S. & Allen, C.R. (1974) Potentially active faults in dam foundations. *Geotechnique*, 24(3), 367–428.

Terzaghi, K. (1946) *Rock Defects and Loads on Tunnel Supports*. Harvard University.

Varnes, D. (1978) Special report, chap. Slope movement types and processes, *Landslides-analysis and control: National Research Council*. Transportation Research Board, Washington, DC., pp.13–33.

Verhoogen, J., Turner, F.J., Weiss, L.E., Wahrhaftig, C. & Fyfe, W.S. (1970) *The Earth*, Holt, Rinehart and Winston, New York. p.748.

Wyllie, D.C. & Mah, C. (2014) *Rock Slope Engineering*. CRC Press, London.

Index